Environmental Science, Engineering and Technology Series

ENVIRONMENTAL EFFECTS OF OFF-HIGHWAY VEHICLES

Environmental Science, Engineering and Technology Series

Nitrous Oxide Emissions Research Progress
Adam I. Sheldon and Edward P. Barnhart (Editors)
2009 ISBN: 978-1-60692-267-5

The Amazon Gold Rush and Environmental Mercury Contamination
Daniel Marcos Bonotto and Ene Glória da Silveira
2009 ISBN: 978-1-60741-609-8

Environmental Effects of Off-Highway Vehicles
Douglas S. Ouren, Christopher Haas, Cynthia P. Melcher, Susan C. Stewart, Phadrea D. Ponds, Natalie R. Sexton, Lucy Burris, Tammy Fancher and Zachary H. Bowen
2009 ISBN: 978-1-60692-936-0

Environmental Science, Engineering and Technology Series

Environmental Effects of Off-Highway Vehicles

Douglas S. Ouren
Christopher Haas
Cynthia P. Melcher
Susan C. Stewart
Phadrea D. Ponds
Natalie R. Sexton
Lucy Burris
Tammy Fancher
and
Zachary H. Bowen

Nova Science Publishers, Inc.
New York

For permission to use material from this book please contact us:
Telephone 631-231-7269; Fax 631-231-8175
Web Site: http://www.novapublishers.com

LIBRARY OF CONGRESS CATALOGING-IN-PUBLICATION DATA
Available upon request

ISBN 978-1-60692-936-0

Published by Nova Science Publishers, Inc. ✣ New York

CONTENTS

PREFACE

This book and its associated appendixes compile and synthesize the results of a comprehensive literature and Internet search. The literature search was undertaken to uncover information regarding the effects of off-highway vehicle (OHV) use on land health, or "natural resource attributes," and included databases archiving information from before OHVs came into existence to May 2006. Information pertaining to socioeconomic implications of OHV activities is included as well. The literature and Internet searches yielded approximately 700 peer-reviewed papers, magazine articles, agency and non-governmental reports, and internet websites regarding effects of OHV use as they relate to the Bureau of Land Management's (BLM) standards of land health. Discussions regarding OHV effects are followed by brief syntheses of potential indicators of OHV effects, as well as OHV-effects mitigation, site-restoration techniques, and research needs.

This is an edited, reformatted and augmented edition of a USGS Science for a Changing World publication, Open-File Report 2007-1353.

ACKNOWLEDGMENTS

This project was funded by the U.S. Bureau of Land Management National Science and Technology Center, through Project Officer Victoria "Vicki" Josupait, as well as Travis Haby and Charisse Sydoriak. The authors would like to thank the following individuals for the time they offered in review, comment, and development of this document: Peter Doran, Travis Haby, Jeffery E. Herrick, Victoria "Vicki" Josupait, Mike "Sherm" Karl, David A. Pyke, James E. Roelle, Robert Stottlemyer, and Charisse Sydoriak.

Conversion Factors

Inch/Pound to SI (International System of Units)

Multiply	**By**	**To obtain**
Length		
inch (in.)	2.54	centimeter (cm)
inch (in.)	25.4	millimeter (mm)
foot (ft)	0.3048	meter (m)
mile (mi)	1.609	kilometer (km)
yard (yd)	0.9144	meter (m)
Area		
square foot (ft_2)	929.0	square centimeter (cm_2)
square foot (ft_2)	0.09290	square meter (m_2)
square inch (in_2)	6.452	square centimeter (cm_2)
square mile (mi_2)	2.590	square kilometer (km_2)
Volume		
cubic inch (in_3)	16.39	cubic centimeter (cm_3)
cubic yard (yd_3)	0.7646	cubic meter (m_3)
Mass		
ounce, avoirdupois (oz)	28.35	gram (g)
pound, avoirdupois (lb)	0.4536	kilogram (kg)
ton, short (2,000 lb)	0.9072	megagram (Mg)

SI to Inch/Pound (English System of Units)

Multiply	**By**	**To obtain**
Length		
centimeter (cm)	0.3937	inch (in.)
millimeter (mm)	0.03937	inch (in.)
meter (m)	3.281	foot (ft)
kilometer (km)	0.6214	mile (mi)
meter (m)	1.094	yard (yd)

Table. Continued

Multiply	By	To obtain
Area		
square centimeter (cm^2)	0.001076	square foot (ft^2)
square meter (m^2)	10.76	square foot (ft^2)
square centimeter (cm^2)	0.1550	square inch (in^2)
square kilometer (km^2)	0.3861	square mile (mi^2)
Volume		
cubic centimeter (cm^3)	0.06102	cubic inch (in^3)
cubic meter (m^3)	1.308	cubic yard (yd^3)
Mass		
gram (g)	0.03527	ounce, avoirdupois (oz)
kilogram (kg)	2.205	Pound, avoirdupois (lb)
megagram (Mg)	1.102	ton, short (2,000 lb)
megagram (Mg)	0.9842	ton, long (2,240 lb)

GLOSSARY

ATV (All-Terrain Vehicle) Small, motorized 3- or 4-wheeled vehicles specifically designed for off-road use. The American National Standards Institute (ANSI) further defines an ATV as a vehicle that travels on low-pressure tires, with a seat that is straddled by the operator, and with handlebars for steering control. By the current ANSI definition, it is intended for use by a single operator, although a change to include 2-seaters (in tandem) is under consideration. Herein, the definition of ATV coincides with the description above and does not include passenger vehicles, including sport-utility vehicles or 4-wheel-drive jeeps.

fugitive dust Dust raised by mechanical (anthropogenic) disturbance of granular material exposed to and becoming suspended in the air, then carried by wind. Arises from "nonpoint" sources—such as unpaved roads, agricultural tilling operations, aggregate storage piles, and heavy construction—rather than "point" sources—such as confined flow streams discharged to the atmosphere from a stack, vent, or pipe.

indicator threshold For a given land health indicator (or set of indicators), the value(s) at or above which management action may be triggered or required.

land health The condition of natural resource attributes, including soils and site stability, hydrologic function, and biotic integrity.

OHV Defined herein as any civilian off-highway vehicle, including motorcycles, motorized dirt bikes, ATVs (see definition above), snowmobiles, dune buggies, 4-wheel-drive jeeps, sport-utility vehicles, and any other civilian vehicles capable of off-highway, terrestrial travel (including utility vehicles [UTVs] and ATVs with more than 4 wheels).

OHV route Defined herein as any unpaved route created for OHV travel, including single-track paths or trails, two-tracks, and unimproved or improved dirt/gravel roads. Herein, this term is also applied to "rogue" (undesignated or unauthorized) routes created by OHV users in closed or limited areas.

population dynamics	Herein, used broadly to include wildlife or vegetation population size, density, and/or distribution (both spatial and temporal); rates of birth/germination, death, and/or survivorship; population gender/age-class structure; population genetics; and/or the rates/directions of change in all these parameters.
right-of-way habitat	Habitat provided within the legal description of a given transportation corridor.
sink population	For a given metapopulation, a population sink is a local area or habitat where the local population's reproductive rate is lower than the required replacement rate (in other words, a sink population is eventually extirpated without immigration of individuals from other areas). Population sinks often occur where there is excessive predation pressure and/or poor habitat quality.
source population	For a given metapopulation, a population source is a local area or habitat where the local population's reproductive rate is greater than the required replacement rate. Excess individuals produced from a source population may emigrate to join sink populations, thereby keeping the sink populations from becoming extirpated.
stream order	A stream's order is determined by its confluence with other streams: first-order streams are headwaters, the confluence of two first-order streams forms a second-order stream, the confluence of two second-order streams forms a third-order stream, and so on.

Executive Summary

This report and its associated appendixes compile and synthesize the results of a comprehensive literature and Internet search conducted in May 2006. The literature search was undertaken to uncover information regarding the effects of off-highway vehicle (OHV) use on land health, or "natural resource attributes," and included databases archiving information from before OHVs came into existence to May 2006. Information pertaining to socioeconomic implications of OHV activities is included as well. The literature and Internet searches yielded approximately 700 peer-reviewed papers, magazine articles, agency and non-governmental reports, and internet websites regarding effects of OHV use as they relate to the Bureau of Land Management's (BLM) standards of land health. Discussions regarding OHV effects are followed by brief syntheses of potential indicators of OHV effects, as well as OHV-effects mitigation, site-restoration techniques, and research needs.

Terminology Used in this Report

The BLM has definitions for several road and trail types; however, the OHV literature often uses somewhat different definitions. Whereas all terms are useful within their own contexts, herein the general term "OHV routes" is used to simplify discussions concerning all types of unpaved roads and trails, whether designated or unauthorized, and "roads" or "highways" are used to simplify discussions concerning paved roads. The definition of OHV also varies by agency and author, and to simplify discussions herein, OHV may include off-highway motorbikes, ATVs, dune buggies, snowmobiles, 4-wheel drive jeeps, motorcycles, some types of 4-wheel drive automobiles (including sport-utility vehicles), and any other civilian vehicle specifically designed for off-road travel. OHV type or route/road type are specified if a given discussion warrants and if the literature cited in that discussion specified OHV type or route/road type.

How to Use this Report

Major sections of this document comprise a "manager's report," which includes a literature synthesis and related discussions of (1) OHV effects on natural resource attributes and socioeconomics; (2) indicators described in the literature to evaluate/monitor OHV effects on natural resource attributes and could serve as potential indicators in future research or monitoring programs; (3) mitigation and site-restoration techniques used for OHV-use areas; and (4) research and monitoring needs pertaining to OHV-effects. This document also includes extensive bibliographies pertaining to OHV effects on natural resources. It is recommended that readers focus first on the manager's report, as it provides the basic understanding of OHV effects and potential approaches to researching, monitoring, and/or managing OHV effects. Reading the Executive Summary, the summaries provided in each section, and the conclusion may suffice for those seeking a quick overview. *To facilitate a rapid review, section summaries are placed at the beginning of their respective sections and do not contain in-text citations. For a more in-depth review (with in-text citations), the entire manager's report should be read.* Appendix 1 provides the extensive bibliographies, and Appendix 2 details the literature/Internet search methods and summarizes the search results (including tables and graphs).

OHV Effects

OHV Effects on Soils and Watersheds

The primary effects of OHV activity on soils and overall watershed function include altered soil structure (soil compaction in particular), destruction of soil crusts (biotic and abiotic) and desert pavement (fine gravel surfaces) that would otherwise stabilize soils, and soil erosion. Indicators of soil compaction discussed in the OHV effects literature include soil bulk density (weight per unit of volume), soil strength (the soil's resistance to deforming forces), and soil permeability (the rate at which water or air infiltrate soil). Generally, soil bulk density and strength increase with compaction, whereas permeability decreases with compaction. As soil compaction increases, the soil's ability to support vegetation diminishes because the resulting increases in soil strength and changes in soil structure (loss of porosity) inhibit the growth of root systems and reduce infiltration of water. As vegetative cover, water infiltration, and soil stabilizing crusts are diminished or disrupted, the precipitation runoff rates increase, further accelerating rates of soil erosion.

OHV Effects on Vegetation

Plants are affected by OHV activities in several ways. As implied above, soil compaction affects plant growth by reducing moisture availability and precluding adequate taproot penetration to deeper soil horizons. In turn, the size and abundance of native plants may be reduced. Above-ground portions of plants also may be reduced through breakage or crushing, potentially leading to reductions in photosynthetic capacity, poor reproduction, and diminished litter cover. Likewise, blankets of fugitive dust raised by OHV traffic can disrupt photosynthetic processes, thereby suppressing plant growth and vigor, especially along OHV routes. In turn, reduced vegetation cover may permit invasive and/or non-native plants—particularly shallow- rooted annual grasses and early successional species capable of rapid establishment and growth—to spread and dominate the plant community, thus diminishing overall endemic biodiversity.

OHV Effects on Wildlife and Habitats: Native, Threatened, and Endangered Species

Habitats for native plants and animals, including endangered and threatened species, are impacted by OHVs in several ways. A salient effect is habitat fragmentation and reduced habitat connectivity as OHV roads and trails proliferate across the landscape. Reduced habitat connectivity may disrupt plant and animal movement and dispersal, resulting in altered population dynamics and reduced potential for recolonization if a species is extirpated from a given habitat fragment. Wildlife is also directly affected by excessive noise (decibel levels/noise durations well above those of typical background noise) and other perturbations associated with OHV activities. Disturbance effects range from physiological impacts—including stress and mortality due to breakage of nest-supporting vegetation, collapsed burrows, inner ear bleeding, and vehicle-animal collisions—to altered behaviors and population distribution/dispersal patterns, which can lead to declines in local population size, survivorship, and productivity.

OHV Effects on Water Quality

The effects of OHV activities on water quality can include sedimentation (deposited solids), turbidity (suspended solids), and pollutants within affected watersheds. Sedimentation increases because compacted soils, disrupted soil crusts, and reduced vegetation cover can lead to increased amounts and velocities of runoff; in turn, this accelerates the rates at which sediments and other debris are eroded from OHW-use areas and flushed to aquatic systems downslope. Pollutants associated with deposition of OHW emissions and spills of petroleum products may be adsorbed to sediments, absorbed by plant material, or dissolved in runoff; once mobilized, these contaminants may enter aquatic systems.

OHV Effects on Air Quality

Air quality is affected when OHW traffic raises fugitive dust and emits by-products of combustion. Because wind can disperse suspended particulates over long distances, dust raised by OHW traffic can blanket plant foliage and disperse dust-adsorbed contaminants well beyond a given OHW-use area. Primary combustion by-products potentially affecting air quality in OHWuse areas include (but are not limited to) polycyclic aromatic hydrocarbons, sulfur dioxide (SO2), nitrogen oxides (NO_x), and ozone (O3). Although leaded gasoline has not been used in the United States since 1996, lead emissions deposited prior to the ban on leaded gasoline may persist for decades and continue impacting ecosystems as wind and water erosion continue to mobilize lead and other contaminants downwind (or downslope) of contaminated soils.

Socioeconomic Implications of OHV Use

For the purposes of this document, the socioeconomics of OHW use include (1) OHW user demands, concerns, and attitudes; (2) the economic effects of OHW use on communities near OHW-use areas; (3) and the effects of OHW use on other land users. Although not one of BLM's land health considerations, the socioeconomic implications of OHW use have significant direct and indirect effects on land health. As the popularity of OHW recreation increases, socioeconomic factors become increasingly important considerations in understanding and mitigating the overall effects of OHW use on land health. OHW recreation can have significant economic value to local communities where and when OHW use is popular; however, the economic costs to those communities remain unknown. OHW use also can lead to conflicts among different land users—both OHW users and people seeking non-motorized forms of recreation—within OHW-use areas and nearby areas. Crowding of designated OHW areas may encourage unauthorized use in closed areas, and adjacent or overlapping use types may cause dissatisfaction or discourage recreation altogether, which can diminish public support for land- management programs.

Potential Indicators (Both Direct and Indirect) for Evaluating and Monitoring OHV Effects

Soil Health and Watershed Condition

- Soil strength
- Soil bulk density
- Soil permeability (rates of air and water infiltration)
- Erosion rate
- Level of sedimentation or turbidity in wetlands
- Surface changes (for example, gully erosion)
- Presence/condition of soil crusts

Vegetation Health

- Plant community composition (including species and structural diversity, ratio of native to non-native or invasive species)
- Abundance of individuals and/or stem density
- Percent vegetation cover
- Plant size
- Growth rate
- Biomass

Habitat Condition and Health of Wildlife Populations (including indirect indicators)

- Habitat patch size and connectivity
- Community composition (including species diversity, ratio of native to non-native or invasive species)
- Population size, density, and trend
- Spatiotemporal distribution of populations
- Survivorship and mortality rates
- Productivity and body mass
- Age-class and gender structure
- Frequency of OHVs passing through a given area and associated wildlife mortalities rates
- Road or trail type and width
- Level (decibels), duration, and timing of traffic noise

Water Quality

- Sedimentation rate
- Levels of turbidity and suspended solids
- Contaminant levels, including petroleum-derived compounds from spills and emissions, such as benzene; ethylbenzene; m-, p-, and o-xylene; toluene; 1,3-butadiene; and lead

Air Quality

- Level of dust particulates
- Particulate levels of OHV emission by-products, such as polycyclic aromatic hydrocarbons, aldehydes, carbon monoxide, nitrogen oxides, ozone, and sulfur oxides

Socioeconomics

- User satisfaction with recreation experiences
- User compliance with OHV (or other) regulations
- User knowledge regarding effects of recreation activities on various aspects of land health
- Distribution and intensity of OHV versus non-motorized recreation and other land uses
- Extent to which unauthorized trails are created and damage to vegetation occurs
- Trends in local economic indicators associated with OHV and non- motorized recreation and other land uses, such as sales of camping equipment, gasoline, restaurants, lodging facilities

Mitigation and Site-Restoration Techniques

Balancing OHV-user preferences with protecting land health and the needs of other land users requires careful study and planning, as well as appropriate management strategies. Prior planning for locating OHV areas before they are opened to the public can preclude undesirable effects of OHV use and costly site restoration. Once a site has been used, however, trail/area closures, signage, and other visual cues, as well as enforcement and limiting visitor numbers through "rationing," are among the tools used to preclude additional effects.

Because habitat fragmentation is particularly difficult to repair, planning and management designed to maintain habitat connectivity are crucial to minimize fragmentation. Variation in impacts requires various restoration techniques. Whereas a single OHV pass on a xeric landscape may cause long-lasting damage, a similar single pass on a mesic landscape may require no treatment at all. Restoration approaches may include replacing native soil where erosion has removed the topsoil and exposed the underlying bedrock, seeding with indigenous plants, inoculating soils with native microbes and mycorrhizae, scarifying, and/or mulching. Ultimately, the success of such measures depends on the nature and intensity of the disturbance, topography, soil type, climate, and the ability of land managers to enforce closures and prevent the proliferation of new routes.

Monitoring and Research Needs to Support OHV Management Decisions

Elucidating OHV impacts on soils and watersheds, vegetation, wildlife and their habitats, water and air quality, and/or socioeconomics, whether through monitoring or experimental research studies, will require careful planning and appropriate, rigorous study design. Results of many past ecological studies on the effects of OHVs may be regarded as preliminary, particularly for those that lacked comparable treatment and control sites and site replication. Overall, the reliability and value of monitoring and research results would increase significantly by including a broad range of spatial and temporal scales—from microhabitat to landscape or ecosystem scale, and from short-term (seasonal) to long-term (decades)—within the full range of impacted habitat types represented on BLM lands. The full array of site types also needs evaluation, including designated OHV-use sites, undesignated (rogue) OHV-use sites, unused areas, and restoration sites. Multiple- and simultaneous-assessment techniques that take advantage (and push the advancement) of existing and emerging technologies are needed to fully represent the scale and diversity of OHV impacts on the abiotic and biotic components of affected lands and communities, and to ascertain indicator thresholds as they pertain to BLM's land health standards. More specifically, monitoring and research needs pertaining to OHV use and impacts include (but are not limited to)

- well-designed monitoring programs and experimental studies that incorporate planned comparisons of treatment (OHV-impacted) and control (unimpacted/ reference) sites;
- before and after OHV-impact studies;
- studies at various spatial and temporal scales across all impacted habitat types;
- studies on habitat fragmentation and road-edge effects caused by OHV activities;
- studies on various gradients in OHV disturbance at various distances from OHV routes;
- studies to improve the understanding of the physical and chemical dynamics of soil compaction;

- studies evaluating the effects of erosion, sedimentation, and turbidity at both local (immediately downslope of OHV-affected sites) and landscape (throughout impacted watersheds) scales;
- studies of OHV effects on plant and animal population dynamics;
- simultaneous evaluations of wildlife responses and OHV route-specific variables;
- improvements in techniques for successful site restoration;

- improvements in techniques and technologies for assessing OHV impacts over large areas and long periods of time;
- effectiveness evaluations of various techniques to manage OHV use and its ecological and socioeconomic effects while simultaneously providing the greatest satisfaction among all land users; and
- studies that determine the economic and sociological costs of OHV use.

Chapter 1

INTRODUCTION

1.1. ISSUE CONTEXT: BUREAU OF LAND MANAGEMENT LAND HEALTH AND OFF-HIGHWAY VEHICLE USE

1.1.1. Bureau of Land Management Land Health Standards

In the mid-1990s, *Rangeland Health: New Methods to Classify, Inventory, and Monitor Rangelands* was published to examine the scientific basis and success of methods used by federal agencies to inventory, classify, and monitor rangelands, and to make recommendations for improvements to these methods (Committee on Rangeland Classification, Board on Agriculture, and National Research Council, 1994). Therein, rangeland health was defined as "...the degree to which the integrity of the soil and the ecological processes of rangeland ecosystems are sustained." In 2001, the BLM published *Rangeland Health Standards*, a process framework for assessing rangeland health via interdisciplinary teams that, at a minimum, would evaluate (1) watershed function; (2) nutrient cycling and energy flow; (3) water quality; (4) habitat for endangered, threatened, proposed, candidate, or special status species; and (5) habitat quality for native plant and animal populations (U.S. Bureau of Land Management, 2001). Subsequently, it was clarified that the term "rangeland" is interchangeable with "land" and that "... 'the rangeland health standards' really apply to the condition of the land itself regardless of the uses that may influence the health of that land" (U. S. Bureau of Land Management, 2007). In other words, the standards for rangeland health apply to all BLM lands, whether used for grazing, off-highway vehicle (OHV) recreation, or any other use permitted to occur on BLM lands.

The standards of rangeland health establish minimum resource conditions that must be achieved and maintained to ensure the "proper functioning condition" (see Barrett and others, 1995)—both physical and biological—and sustainability of BLM lands. These standards, however, must be relative to the native conditions for a given "reference site," as native conditions vary widely among sites on BLM lands. For example, the BLM defines healthy soils in northwest California as "exhibit[ing] characteristics of infiltration, fertility, permeability rates, and other functional and physical characteristics that are appropriate to soil type, climate, desired plant community, and land form" (personal communication from M. Karl to V. Josupait, U.S. Bureau of Land Management, Denver, Colorado, May 2001).

The BLM's technical reference on *Interpreting Indicators of Rangeland Health* recognizes three broad categories of natural resource attributes for assessing land health: (1) soils and site stability, (2) hydrologic function, and (3) biotic integrity (Pellant and others, 2005). Relative to off-highway vehicle (OHV) impacts on ecosystem health, soil/site stability and hydrologic or watershed function pertain (but are not limited) to erosion and extent of surface changes and patterns of water flow, including infiltration. Biotic integrity pertains to community structure and functionality of plants. Within these three categories, overall land health is qualitatively assessed via 17 parameters (or indicators) that indicate the presence, number, extent, percent, and/or depth or height of (1) rills, (2) water flow patterns, (3) erosional pedestals/terracettes, (4) bare ground, (5) gullies and gully erosion, (6) wind scoured blowouts and/or depositional areas, (7) litter movement, (8) soil surface resistance to erosion, (9) soil surface structure and content of soil organic matter (SOM), (10) effect of plant community composition and spatial distribution on infiltration and runoff, (11) compaction layer, (12) dominance hierarchy of functional/structural groups in plant communities, (13) mortality and decadence among plant functional groups, (14) litter cover, (15) expected above-ground annual production, (16) potential invasive species, and (17) perennial plant reproductive capability (Pellant and others, 2005). Many of the indicators listed above may be assessed in terms of the condition and extent of abiotic (chemical) and biotic soil crusts, and soil surfaces of small stones known as "desert pavement," as they help stabilize soils and/or cycle nutrients through the system. Indicators 1-11 largely pertain to soil/site stability and hydrologic/watershed function, whereas parameters 12-17 are primarily indicators of biotic integrity, although there is overlap among the two groups.

1.1.2. Increasing OHV Use

An important factor affecting the health of BLM lands is the use of OHVs. In 1993, 2,920,000 all-terrain vehicles (ATVs) and off-highway motorcycles were estimated to be in use (Cordell and others, 2005). Between 1995 and 2003, sales of ATVs and off-highway motorcycles tripled, increasing the number of ATVs and off-highway motorcycles in use to 8,010,000 (Cordell and others, 2005). Because the popularity of OHV-based recreation is relatively recent and still increasing (see Matchett and others, 2004), the full range of short- and long-term impacts has yet to be fully realized or understood. Overall, it is clear that OHV use on public lands is and will continue to be an important management issue.

1.2. Objectives, Scope, Organization, and Use of this Report

1.2.1. Objectives

The objectives of this report are twofold. This first is to synthesize the results of a comprehensive literature search on what is currently known about the effects of OHV activities as they relate to the BLM's land health standards (U.S. Bureau of Land Management, 2001). These discussions include socioeconomic implications of OHV use—including preferences of OHV users, effects of OHV activities on other land users, and the economic impacts of OHV recreation on local economies—because understanding these factors and incorporating that knowledge into management plans and policies will be crucial to management success. The second objective is to discuss the indicators of land health and socioeconomics described in the OHV effects literature, as they have potential usefulness for evaluating or monitoring lands and land users affected by OHV use. This report also contains brief overviews of mitigation approaches, site-restoration techniques, and monitoring and research needs described in the OHV-effects literature.

1.2.2. Geographical Scope

Although the vast majority of literature and other sources consulted address OHV impacts on ecosystems in the western United States, there are a number of useful references that address OHV use or impacts of roads in other parts of the United States and in other countries. Some of these sources have been included to provide additional information not provided elsewhere and/or to broaden the scope and relevance of this document.

1.2.3. Organization

The main body of this document—referred to herein as the "manager's report" (Executive Summary and Sections 1-7)—includes the literature synthesis of OHV effects; annotated bibliographies that typify the body of research regarding effects of OHV activities on BLM's land health standards; sections regarding indicators of OHV effects, mitigation and site-restoration techniques, and monitoring and research needs; and a listing of all literature cited in the manager's report. To facilitate the logical flow of information, the natural resource attributes addressed in BLM's land health standards serve as the underlying organizational structure of all major sections in the report, as indicated by parallel subsection headings pertaining to (1) soils and watersheds, (2) vegetation, (3) habitat for native plants and wildlife, (4) water quality, and (5) air quality. Although not part of the BLM's land health agenda, a sixth subsection has been included in each major section to address the socioeconomic considerations of OHV use, because, ultimately, it is socioeconomic factors (including conflicts among land users, preferences of OHV users, and economics of OHV use) that drive changes in overall land health.

1.2.4. Tips on Navigating this Document

It is recommended that readers focus first on the manager's report, as it provides the basic understanding of OHV effects and possible indicators to use in research, monitoring, and/or management programs. Reading the Executive Summary, subsection summaries, and the overall conclusion may suffice for those seeking a quick overview. *For quick and easy reference, subsection summaries on OHV effects are placed at the top of their respective subsections. In-text citations were purposely left out of all summaries to enhance readability. For a more in-depth, fully cited review, reading the entire manager's report is recommended.*

Appendixes 1 and 2 provide further information and more extensive resources for those needing high levels of detail. The Extensive Bibliographies (Appendix 1) includes the approximately 700 publications and reports uncovered through the literature search (from 1960 to May 2006) that pertain to effects of OHV use. Appendix 2 details the literature- and Internet- search methods and provides tabular and graphical summaries of literature, websites, and associated resources, as they pertain to OHV effects on natural resources and related policies.

1.3. Definitions of OHV Routes/Roads, Vehicles, and Activities

1.3.1. Definitions of Roads and Trails Used in this Report

BLM's *Roads and Trails Terminology* document (U.S. Bureau of Land Management, 2006) defines (1) a road as "a linear route declared a road by the owner, managed for use by low- clearance vehicles having four or more wheels, and maintained for regular and continuous use;" (2) a primitive road as "a linear route managed for use by four-wheel drive or high-clearance vehicles;" and (3) a trail as "a linear route managed for human-powered, stock, or off-highway vehicle forms of transportation or for historical or heritage values." Bolling and Walker (2000) offer more detailed definitions of unpaved roads and trails: (1) graded, improved roads are those from which the topsoil has been removed by bulldozer and characterized by the presence of lateral berms; (2) unimproved roads are those not graded consistently; (3) jeep trails are four- wheel drive tracks impacted only by vehicular traffic and generally characterized by a center berm; and (4) single tracks are severely compacted trails generated by OHVs.

The literature includes numerous reviews on the ecological effects of paved as well as unpaved roads (see Andrews, 1990; Forman and Alexander, 1998; Spellerberg, 1998; Trombulak and Fris sell, 2000). Ecosystems of the West, however, are especially vulnerable to OHV-related activities on unpaved (gravel or dirt) roads and trails due to the effects they impose on soils and vegetation, which may take centuries to recover (Webb, 1982; Lovich and Bainbridge, 1999). Furthermore, unpaved roads comprise the majority of OHV routes used throughout public lands in the western United States. Therefore, the primary considerations in this report are unpaved roads and trails. References and discussions regarding effects of paved roads are not entirely excluded, however, as they often inform the potential scope of OHV effects not otherwise addressed. When necessary to do so, the type of road is specified.

All of terms described above for roads and trails are useful within their own contexts, but to distinguish between them and those used in the OHV literature would unnecessarily complicate discussions herein. Therefore, except where there is a need to specify in more detail, unpaved roads, primitive roads, and unpaved trails, are referred to as "routes," regardless of their intended purpose or how they are maintained. "Routes" also include unauthorized or "rogue" roads and trails created by OHV users traveling off officially designated roads, primitive roads, and trails. Paved roads are referred to as "roads."

1.3.2. Definitions of OHVs and OHV Activities Used in this Report

BLM's *Roads and Trails Terminology* document (U.S. Bureau of Land Management, 2006) defines an OHV as "any motorized vehicle capable of—or designated for—travel on or immediately over land, water, or other natural terrain" (excluding nonamphibious registered motorboats; military, fire, emergency, or law enforcement vehicles used for emergency purposes; official vehicles used expressly by an authorized officer; and military vehicles). Cordell and others (2005) further specify that OHVs may include motorcycles and off-highway motorbikes, ATVs, dune buggies, snowmobiles, most 4-wheel drive automobiles (jeeps, sport utility vehicles), and any other civilian vehicle specifically designed for off-road travel. For the purpose of this document, OHV is defined in accordance with BLM terminology and includes those vehicles listed by Cordell and others (2005), as well as utility vehicles (UTVs) and ATVs with more than 4 wheels.

There are numerous activities and outcomes directly and indirectly associated with OHV use. To simplify discussions of OHV activities herein, the term "OHV activities" largely refers to driving OHVs for recreation. In certain contexts, OHV activities also may include driving and parking vehicles that tow trailers carrying OHVs and loading and unloading OHVs from trailers. Whereas the use of 4-wheel drive jeeps, automobiles, and sport utility vehicles is largely restricted to unpaved roads and jeep trails, the effect of ATVs and off-highway motorcycles extends well beyond them to double- and single-track trails, as well as unauthorized roads and trails. Thus, "OHV activities" may include driving OHVs on authorized roads/routes and on (or creating) unauthorized routes.

Chapter 2

Effects of OHV Travel on Natural Resource Attributes and Socioeconomics

2.1. Scale and Patterns of OHV Activities and their Effects

Temporal and spatial scales are crucial considerations when evaluating or monitoring effects of any factor on ecosystems (Noon, 2003; Ringold and others, 2003). In discussing OHV effects on desert ecosystems, Brooks and Lair (2005) and Matchett and others (2004) describe the impacts of OHV activities at various spatial and temporal scales. At the highly localized spatial scale, one might find soil compaction taking place within the confines of a single OHV route, the effects of which might be limited to poor infiltration of water and reduced plant cover in the route itself. Brooks and Lair (2005) go on to explain, however, that the cumulative impacts of any one effect at many sites can result in impacts at much greater scales. For example, if networks of OHV routes criss-cross large areas, the habitat connectivity that previously facilitated animal movements within that landscape may be disrupted (Forman and others, 2003: p. 129-134). Similarly, a single pass by one OHV probably has negligible effects on animal distributions, but if OHV traffic is intense and chronic, animal densities may decline (Reijnen and others, 1995, 1997) as cumulative impacts of this one effect occurring at many sites across a landscape disrupt entire populations. Furthermore, any *direct* effect of OHV use is also likely to have *indirect* effects that go beyond the site of disturbance. For example, reduced plant cover (direct effect) can result in greater rates of erosion in and around an OHV route, which, in turn, might increase sedimentation and turbidity in wetlands downslope of the route (indirect effect). Overall, Brooks and Lair (2005) conclude that effects of OHV use need to be evaluated at appropriate scales, which must take into account the scale at which OHV activities and ecosystem responses occur (Brooks and Lair, 2005). Overall, most, if not all, effects of OHV activities described in sections 2.2 through 2.6 can occur from the very localized and/or ephemeral scale to the landscape and/or long-term scale. By the same token, any direct effect may have a number of indirect effects, the magnitude of which may depend on the spatial and temporal scales at which a direct effect occurs.

At an OHV site in California, Matchett and others (2004) classified OHV routes in terms that describe intensity and pattern of use. First, OHV lines (routes) were categorized as either *dirt* (lines most likely created for or by OHV use) or as *wash* (lines most likely created by water flow, but possibly used by OHVs). They went on to define levels of OHV use that also imply patterns of use: (1) densely tracked reticulate (OHV lines evident in a web-like pattern, but too dense and overlapping to distinguish individually); (2) densely tracked hill-climb (OHV lines evident on slopes, but too dense and overlapping to distinguish individually; (3) densely tracked intersection (OHV lines evident at intersections, but too dense and overlapping to distinguish individually; (4) densely tracked right-of-way (OHV lines evident near pipelines, transmission lines, and highway right-of-ways, but too dense and overlapping to distinguish individually; (5) densely tracked wash (OHV lines evident within washes, but too dense and overlapping to distinguish individually; (6) denuded hill-climb (OHV lines not readily evident on hill-climbs, but the preponderance of densely tracked areas in the vicinity indicate that OHV use was probably high within that area); and (7) denuded staging (OHV routes not readily evident in relatively flat camping areas, but the preponderance of densely tracked areas in the vicinity indicated that OHV use was probably high within that area) (Matchett and others, 2004). Overall, these categories indicate that OHV use often entails many criss-crossing routes as opposed to a single route. They also point out that OHV use may be heaviest on slopes, along right-of-ways, in washes, and in the vicinity of camping facilities.

2.2. OHV Effects on Soils and Watersheds

2.2.1. Section Summary

Important effects of OHV activities on soils and watershed function include soil compaction, diminished water infiltration, diminished presence and impaired function of soil stabilizers (biotic and abiotic crusts, desert pavement), and accelerated erosion rates. Compacted soil inhibits infiltration of precipitation. In turn, soil moisture available to vegetation is diminished, volumes and velocities of precipitation runoff increase, and soil erosion accelerates, leading to the formation of gullies and other surface changes. Additionally, soil compaction may inhibit root growth among plants, in which case organic matter, litter, soil fertility, and vegetative cover are diminished, further exacerbating the soil's susceptibility to erosion. Where

biotic and chemical crusts or other soil stabilizers are disturbed or destroyed, soil erosion from water and wind may increase beyond rates found in undisturbed sites with similar soils and conditions; nutrient-cycling processes also are likely to be disrupted, potentially leading to declines in soil fertility.

2.2.2. Soil Compaction and Reduced Water Infiltration

One of the most common and important effects of OHV activities is soil compaction (Liddle, 1997), which diminishes water infiltration, destroys soil stabilizers (biotic and abiotic crusts, desert pavement), and promotes greater rates of erosion from water and wind. In turn, soil moisture available for plant growth is diminished, precipitation runoff increases in volume and velocity, and soil erosion accelerates, which leads to surface changes, including the formation of rills, gullies, terracettes, and pedestals (Webb and others, 1978; Iverson and others, 1981; Webb, 1982; Hinckley and others, 1983; Wilshire, 1 983b). The extent of soil compaction may be measured in terms of soil bulk density, soil strength, and/or permeability. Soil bulk density, calculated as oven-dried soil weight per unit of volume, is typically expressed as g/cm^3 or g/cc. Soil strength, measured as the soil's resistance to deforming forces—or the amount of energy required to break apart aggregates or move implements through the soil—is typically expressed as kg/cm^2 or pounds per square inch (PSI). Soil permeability is the rate at which water (or air) infiltrates the soil, expressed as cm/hr or inches/hr (Leung and Meyer, 2004). Generally, soil bulk density and strength increase with increasing compaction, whereas permeability decreases with increasing compaction (Adams and others, 1982; Webb, 1982; Cole, 1990).

Important factors affecting a soil's susceptibility to compaction include its (1) texture (relative proportions of sand, silt, and clay); (2) structure (the grouping of sand, silt, and clay particles into aggregates), including its porosity (a measure of pore space, which affects the amount of air or water a soil can hold) and aggregate stability (the ability of soil aggregates to resist disruption from outside forces—water in particular); (3) type (series) and depth; and (4) antecedent moisture (the soil's water content prior to compaction). Sandy or clayey soils relatively uniform in texture and structure are less vulnerable to compaction than loamy sands or coarse-textured, gravelly soils characterized by variability in particle size (Lovich and Bainbridge, 1999). In addition, soils with greater water content are more susceptible to compaction than those containing less moisture (Webb, 1982), although even in semi-arid and arid lands soil compaction is problematic because the texture of these soils is slow to recover (Webb, 1982) through natural soil-loosening processes (including shrinking, swelling, drying, wetting, freezing, and thawing).

As the number of vehicle "passes" (one pass is the equivalent of one OHV passing over a given area one time) increases, soil bulk density and soil strength increase and permeability (as indicated by water infiltration rate) decreases (Lovich and Bainbridge, 1999). Soil compaction may become evident after only a few vehicle passes. In fact, Iverson and others (1981) found that soil bulk density increased logarithmically with the number of vehicle passes. Similarly, Adams and others (1982) report that soil strength on routes subjected to a single vehicle pass was 5.3 to 28.4 kg/cm^2 (75.3 66 to 403.848 PSI) greater (depending on the percent soil moisture) than that of nearby undisturbed soils; after 10 to 20 passes, soil strength was too great (impenetrable) to measure with a penetrometer, indicating that a few passes were enough to cause soil "cementation." After initial disturbance, the effects of soil compaction can persist for years, even centuries, before natural soil-loosening processes can restore the soil's texture (Webb and Wilshire, 1980; Webb, 1982; Froehlich and others, 1985; Prose, 1985; Lovich and Bainbridge, 1999). For example, one year after impact, a one-pass trail was still faintly visible, as indicated by slightly more surface gravel and growth of annual plants (the first to grow in disturbed sites) than on surrounding land, and trails impacted by 100 and 200 passes had notable side berms (Prose, 1985).

Other effects of soil compaction include changes to soil horizons and increased compaction in deeper strata. The OHV traffic associated with the annual Johnson Valley-Parker OHV race (1980-1983) near Joshua Tree National Park on the Colorado River compacted 2 to 5 cm (0.8 to 2.0 in) of the underlying vesicular soil horizon (composed of fined-grained, windblown material occurring about 20 cm [7.9 in] deep near surface soil horizons, often immediately under desert pavement, and characterized by small pores, or vesicles, of air space; typical of arid regions) and caused excavation (mechanical erosion) of the A and B soil horizons to depths of 20 cm (7.9 in) (Wilshire, 1 983a). Prose (1985) found that resistance (to a penetrometer) of soil affected by military maneuvers (including tanks, tracked equipment and personnel carriers, and support vehicles) was 50 percent greater than that of undisturbed soils. Overall, traffic typically causes significant changes to soils, which may take years, if not decades, to recover.

2.2.3. Effects on Soil Stabilizers and Rates of Soil Erosion

A significant effect of soil compaction is the soil's inability to support vegetation after disturbance, thus increasing its susceptibility to erosion (Webb and others, 1978). Soil erosion resulting from soil compaction is caused by two main factors (Hinckley and others, 1983): reduced infiltration rates and destruction of soil stabilizers. Infiltration of water into soils depends, to a large extent, on the soil's porosity, which is reduced by compaction. Soil stabilizers, which are characteristic of undisturbed desert substrates, may include cryptobiotic crusts of lichen, fungi, bacteria, mosses, and/or algae; chemical or mechanical crusts (thin upper coating of clay particles oriented parallel to the surface); and desert pavements (closely packed, interlocking fragments of pebble- and/or cobble-sized rocks from which fine-grained materials have been removed by wind or water erosion) (Lovich and Bainbridge, 1999). Cryptobiotic organisms facilitate accumulation of organic materials and nutrients, including nitrogen and carbon, thereby increasing soil fertility (Johansen, 1993). Since they occur in the soil's upper layer, they also promote water infiltration and enhance retention of soil moisture (Belnap and Gardner, 1993). Their proximity to the surface, however, makes them susceptible to destruction by vehicular and foot traffic.

Cole (1990) documented destruction of cryptogamic soil crusts after only 15 passes by hikers wearing lug-soled boots. Traffic from the Johnson Valley-Parker OHV race mentioned in section 2.2.2 not only destroyed the vesicular soil horizon, it destroyed the overlying desert pavement (Wilshire, 1 983a). Webb (1982), who evaluated soil surfaces (shape; another measure of soil compaction) after 1, 10, 100, and 200 motorcycle passes, found changes occurring after the first few passes, although the effects of subsequent passes were more severe due to their cumulative effects: routes subjected to 100 and 200 passes were characterized by berms and lateral edges, and route midlines were 10-30 mm below the level of surrounding undisturbed ground. Once damaged or destroyed, it may take 300-500 years per inch for soil stabilizers to recover or return to their original state (Hudson, 1971).

Typically, undisturbed soil surfaces are very important in controlling the soil's response to precipitation runoff, particularly where the soil surface is covered with fine gravel that overlays soils with large pores (Webb, 1982). In the Mojave Desert, surface runoff was typically five times greater and sediment yield (in runoff) was 10-20 times greater in OHV-impacted areas than in undisturbed areas (Iverson and others, 1981). For various reasons, certain portions of the desert, including dunes, playas, and areas covered with coarse surface material, are fairly resistant to erosion from runoff (Hinckley and others, 1983), whereas vulnerable areas are those where initial infiltration rates are low, slopes are high, and ratios of surface sand/gravel to smaller particles are low (Iverson and others, 1981). The character of precipitation also influences the susceptibility of denuded soil to erosion; erosion rates are typically greater when rainfall events are of long duration and high intensity (Iverson and others, 1981). Disturbed soils also increase the likelihood of debris eroding from areas disturbed by OHV activities (Lovich and Bainbridge, 1999). Indeed, debris flow has been documented to bury plants growing outside the area impacted (Nakata, 1983).

2.2.4. Annotated Bibliography for OHV Effects on Soils and Watersheds

Adams, J.A., Endo, A.S., Stolzy, L.H., Rowlands, P. G., and Johnson, H.B., 1982, Controlled experiments on soil compaction produced by off-road vehicles in the Mojave Desert, California: Journal of Applied Ecology, v. 19, no. 1, p. 167–175.

Under controlled conditions, soil crust properties were measured to determine how rapidly they were altered by passing vehicles. Routes impacted by a single vehicle pass had soil strengths 5.3 to 28.4 kg/cm^2 (75.366 to 403.848 lb/in^2 [or PSI], depending on the percent soil moisture) greater than undisturbed soil, indicating that just a single pass can begin to affect soil strength. Mean soil strength on routes exposed to 10 and 20 passes was too high to measure. Drying caused the soil in the slightly compacted track to become much harder (increased soil strength) than the undisturbed soil.

Belnap, Jayne, 1993, Recovery rates of cryptobiotic crusts—inoculant use and assessment methods: Great Basin Naturalist, v. 53, no. 1, p. 89–95.

Rates of recovery of cyanobacterial-lichen soil crusts from disturbance were examined. Plots were either undisturbed or scalped, and scalped plots were either inoculated with surrounding biological crust material or left to recover naturally. Natural recovery rates were found to be very slow. Inoculation significantly hastened recovery of the cyanobacterial/green algal component, lichen cover, lichen species richness, and moss cover; even with inoculation, however, lichen and moss recovery was minimal.

Belnap, Jayne, 2002, Impacts of off-road vehicles on nitrogen cycles in biological soil crusts— resistance in different U.S. deserts: Journal of Arid Environments, v. 52, no. 2, p. 155–165.

This study was conducted to evaluate short-term impacts of OHVs on lichen cover and the nitrogenase activity (NA) of biological soil crusts on various soil types in the Great Basin, Colorado Plateau, Sonoran, Chihuahuan, and Mojave deserts. Lichen cover was significantly correlated with percent silt in soil (and negatively correlated with percent sand and clay). Disturbance reduced NA at all 26 sites, but significantly at 12; declines were greatest in soils of cooler regions than hotter ones, possibly indicating that non-heterocystic cyanobacterial species are more susceptible to disturbance than heterocystic species. Sandy soils showed greater reduction of NA as sand content increased, while fine-textured soils showed a greater decline as sand content increased. At all sites, higher NA before the disturbance resulted in less impact to NA post-disturbance. These results may be useful in predicting the impacts of off-road vehicles in different regions and different soils.

Cole, D.N., 1990, Trampling disturbance and recovery of cryptogamic soil crusts in Grand Canyon National Park: Great Basin Naturalist, v. 50, no. 4, p. 32 1–325.

Under controlled conditions, cryptogamic soil crusts in Grand Canyon National Park were trampled by hikers to determine how rapidly they were pulverized and how rapidly they recovered. Only 15 passes were required to destroy the structure of the crusts; visual evidence of bacteria and cryptogam cover was reduced to near zero after 50 passes. It took soil crusts one to three years to redevelop, and after 5 years the extensive bacteria and cryptogam cover left little visual evidence of disturbance. Surface irregularity remained low after 5 years, however, suggesting that recovery was incomplete.

Eckert, R.E.J., Wood, M.K., Blackburn, W.H., and Peterson, F.F., 1979, Impacts of off-road vehicles on infiltration and sediment production of two desert soils: Journal of Range Management, v. 32, no. 5, p. 394–397.

This project staged a series of controlled motorcycle and 4-wheel drive vehicle passes, followed by simulated rainfall. Two sites were chosen to represent two different soil types. Infiltration rates were lower and sediment yield was higher after soil was disturbed by vehicular traffic. High sediment yield was attributed to reduced infiltration after 10 minutes; the remaining 20 minutes of the test period were characterized by particles being carried away in runoff water.

Folz, R.B., 2006, Erosion from all terrain vehicle (ATV) trails on National Forest lands (abs.): Proceedings of the 2006 American Society of Agricultural and Biological Engineers, Portland, Oregon, July 9–12, 2006: St. Joseph, Michigan, 2006 American Society of Agricultural and Biological Engineers, *http://asae.frymulti.com/abstract.asp?aid=21056&t=2.*

Concern about unmanaged use of all terrain vehicles (ATV) on U.S. Forest Service lands prompted an experimental study to test the relative effects of low-, medium-, and high-disturbance trails (based on traffic levels), as measured by reduced litter/vegetation and the width and wheel-rut depth of trails. Trail condition was assessed and then subjected to simulated rainfall. A negative relationship between levels of ATV traffic and rainfall infiltration was not statistically significant among disturbance levels; however, there were significant differences in infiltration and measures of erosion between undisturbed and disturbed conditions. Data from this study will be used to estimate ATV traffic-induced erosion and make decisions regarding management of ATV use.

Iverson, R.M., Hinckley, B.S., and Webb, R.M., 1981, Physical effects of vehicular disturbances on arid landscapes: Science, v. 212, no. 4497, p. 915–917.

In 50 rainfall simulation tests, vehicle-use plots had about five times more runoff and 10-20 times greater sediment yield than adjacent unused plots. In a desert environment, such effects may occur even when use of off-road vehicles is light. Recovery times from vehicular traffic were estimated to be nearly 100 years. Erosion rates were calculated from multivariate statistical analyses using 22 experimental factors. The character of the rainfall was identified as the most important variable in predicting increases in erosion.

Sparrow, S.D., Wooding, F.J., and Whiting, E.H., 1978, Effects of off-road vehicle traffic on soils and vegetation in the Denali Highway region of Alaska: Journal of Soil and Water Conservation, v. 33, no. 1, p. 20–27.

This study examined the effects of vehicles on trails. The surface layer of living material was killed on all main trails, although soil morphology was not generally altered except in the surface horizon. Varying amounts of organic matter were lost from the heavily used trails, depending on slope and vehicle type. Soil depth and drainage were the most important factors influencing the condition of the trail. The greatest effects on soils occurred in poorly drained areas or on loose, gravel-free soils that were highly susceptible to erosion.

Tuttle, M., and Griggs, G., 1985, Accelerated soil erosion at three State Vehicular Recreation Areas: central and southern California, *in* Erosion control: A challenge in our time, proceedings of the 16th annual International Erosion Control Association, February 2 1–22, 1985, San Francisco, California: San Francisco, California, International Erosion Control Association, p. 105-115.

Soil erosion rates were evaluated at three State Vehicular Recreation Areas, with a particular focus on hillclimbs. The key factors contributing to erosion rates were slope, length of climb, soil type, and weather. Based on monitoring and catchment basin yield, erosion in open areas dedicated to OHV use was 10 to 25 times greater than in undisturbed areas.

Webb, R.H., 1982, Off-road motorcycle effects on a desert soil: Environmental Conservation, v. 9, no. 3, p. 197–208.

The effects of controlled motorcycle traffic on a Mojave Desert soil in California were studied in order to quantify soil compaction. Four experimental trails treated with 1, 10, 100, and 200 passes with an off-road motorcycle were established in loamy sand at 6.2 percent (by weight) moisture content. Soil penetration resistance, bulk density, infiltration rate, and response to rainfall were measured for undisturbed soil and the experimental trails immediately after the impact, and soil cores were measured in the laboratory to determine pore-size distributions. Soil bulk density was remeasured one year after the impact to ascertain the amount of recovery. The 1-pass trail had a slight surface indentation with knob imprints from the tires. Along the 100- and 200-pass trails, there were berms and lateral edges, and their centers were 10–30 mm below the level of undisturbed soil adjacent to the trail.

Wilshire, H.G., and Nakata, J.K., 1976, Off-road vehicle effects on California's Mojave Desert: California Geology, June 1976, p. 123–132.

This study was designed to evaluate long-term effects of an off-road vehicle race on the desert landscape, in particular the landscape condition after vehicular use (specifically motorcycles), whether or not effects were confined to the areas of direct impact, and how long the physical effects of such activities remained. Visual observations and penetrometer measurements were recorded in five ground types. Soil compaction was the dominant consequence of motorcycle use: penetrometer data revealed decreases in mean penetration depths. Combined with a notable reduction in plant cover, soil compaction significantly increased the potential for erosion. Initial vehicle impact resulted in substantial, immediate mechanical erosion, followed by wind erosion, culminating in the increased potential for water erosion over longer periods of time.

Wilshire, H.G., Nakata, J.K., Shipley, Susan, and Prestegaard, Karen, 1978, Impacts of vehicles on natural terrain at seven sites in the San Francisco Bay area: Environmental Geology, v. 2, no. 5, p. 295–3 19.

Vegetation and soil properties were measured at seven sites exposed to off-road vehicle activities. Impacts on loamy soils included greater soil surface strength and bulk density, lower infiltration rates and soil moisture, extended diurnal temperature ranges, and reduced organic carbon. These effects, combined with the associated loss in vegetative cover, promoted erosion, the rates of which significantly exceeded Federal and local standards, and the increased sediment yield and runoff caused adverse effects on neighboring properties.

2.3. OHV Effects on Vegetation

2.3.1. Section Summary

Relative to plant communities in OHV-impacted areas, those in undisturbed sites are dominated by native plants, invasive species are not increasing, plant growth and reproduction are vigorous, age-classy structures are appropriate to the species, and canopy cover and vertical structure are adequate for dispersing the energy of precipitation runoff and promoting water infiltration. Direct impacts of OHV activities on vegetation include reduced vegetation cover and growth rates, and increased potential for non-native grasses and pioneering species to become established, thus altering vegetation communities. In certain instances, however, the impervious nature of compacted route and paved road surfaces could result in significant runoff that generates greater moisture availability immediately along OHV routes. In turn, this would promote increased vegetation cover and plant abundance than one might find in surrounding areas farther away from OHV routes.

Some important indirect effects of OHV activities on vegetation are tied to soil properties altered by OHV traffic, as soil properties typically influence vegetation growth. OHV roads and trails also create edge habitats, which can generate conditions that promote the encroachment of non-native and invasive plant species. Other indirect effects include increased amounts of airborne pollutants and dust raised by OHV traffic. A blanket of fugitive dust on plant foliage can inhibit plant growth rate, size, and survivorship.

2.3.2. Overall Effects on Vegetation Cover and Community Composition

When soils are severely disturbed, vegetation cover can be reduced significantly (Adams and others, 1982; Prose and others, 1987; Bolling and Walker, 2000) and growth can be impaired (Spencer and Port, 1988; Angold, 1997). As stated in the previous section, even a few passes by vehicles can cause significant changes in soil properties. Adams and others (1982) found reduced cover of desert annuals in tracks created by as few as 1 (on wet loamy sand) to 20 (on dry loamy sand) vehicle passes; the reduction in cover, however, was not due to fewer plants, but to smaller plant sizes. Similarly, Bolling and Walker (2000) found that in OHV routes there were many small individuals of creosote bush (*Larrea tridentata*), but larger plants were few or absent; in control plots, however, there were more large plants and fewer small ones.

Reduced plant sizes are typical where the extent of soil compaction inhibits their roots from penetrating to deeper soil levels. In fact, Adams and others (1982) determined that root growth is precluded at soil strengths of about 20 kg/cm^2 (284.4 lb/in^2). Within tracks made by 1, 3, 10, and 20 vehicle passes, Adams and others (1982) found that annuals with large taproots (for example, pincushion flower [*Chaenactis fremontii*]) decreased, whereas there was significantly greater cover of common Mediterranean grass (*Schismus barbatus*), a non-native grass with a fibrous root system. The fibrous root system of plants that characterized by single cotyledons, such as common Mediterranean grass, allows for easier germination and root growth than is possible for taprooted dicotyledons.

Soil compaction also increases the potential for invasive, non-native annuals and other early successional plants to establish rapidly in OHV routes, whereas native perennials may require at least 5 years to become established (Adams and others, 1982; Prose and others, 1987;

Lovich and Bainbridge, 1999). This is due, in part, to the increased surface moisture availability within the tracks of OHV routes after compaction has reduced the rate of water infiltration, which may favor the rapid germination and growth of non-native and invasive annuals (Adams and others, 1982). In disturbed areas, pioneering species, such as burrobush (*Ambrosia dumosa* and *Hymenoclea salsola*—now *Ambrosia salsola*; see *http://ucjeps.berkeley.edu/cgibin/get_cpn.pl?3578),* often dominate the plant community and typically their percent cover is similar to, or greater than, that of undisturbed areas (Prose and others, 1987). Davidson and Fox (1974) also found that non-native, early-successional species, such as redstem stork's bill (*Erodium cicutarium*) and common Mediterranean grass were common at sites disturbed by OHVs. When comparing vegetation in disturbed versus protected plots, Brooks (1995) found that common Mediterranean grass was the only species with greater biomass in the disturbed plots.

OHV traffic also causes direct impacts to vegetation structures (breakage, smashing), although population-level effects may be difficult to discern in the short term. Overall, the extent of immediate effects increases with the frequency of OHV passes. For example, Webb (1983) found that after a single pass, annual plants on an OHV route remained intact, but most were destroyed after 10 passes. Likewise, a series of studies to evaluate the impacts of OHV traffic on the Federally listed Peirson' s milkvetch (*Astragalus magdalenae peirsonii*) indicated that this plant was more likely to occur at sites closed to OHV activity than at OHV sites that have been rested from OHV activity (Groom and others, 2005); however, additional study indicated that the number of reproducing plants (and the seedbank) was adequate to maintain the milkvetch population (Phillips and Kennedy, 2006; for more reports, go to *http://www.fws.gov/carlsbad/PMV_Docs.htm)*. It remains unclear, however, whether research conducted over longer time scales would yield different results.

2.3.3. Edge Effects along OHV Routes

Roads and trails also create edge habitats (Johnson and others, 1975; Vasek and others, 1975; Adams and Geis, 1983; Andrews, 1990; Holzapfel and Schmidt, 1990; Lightfoot and Whitford, 1991; Reed and others, 1996), resulting in a variety of effects, including changes in vegetation and encroachment of non-native and invasive species (Huey, 1941; Lovich and Bainbridge, 1999). As mentioned in section 2.3.2, the impermeable surfaces of roads and OHV routes shed precipitation, thereby increasing overall moisture availability in the immediate vicinity of the road or route. Additionally, the coarse-textured soils typically found in association with paved roads (roadbed materials laid down prior to paving) permit good water infiltration along road edges (Hillel and Tadmor, 1962); similar conditions may occur along improved gravel routes. The increased moisture availability may promote greater plant vigor along roadsides than in surrounding areas (Johnson and others, 1975), and Angold (1997) indicated that such effects may extend as far as 200 m from road edges. Indeed, several studies have shown that there can be more vegetation cover along roadsides and right-of-ways than in adjacent areas (Johnson and others, 1975; Vasek and others, 1975; Holzapfel and Schmidt, 1990; Lightfoot and Whitford, 1991). Perennial shrubs, in particular, may grow larger and attain greater vigor and density along road edges (Johnson and others, 1975; Lightfoot and Whitford, 1991). Likewise, Johnson and others (1975) found that the standing crop (a measure of primary productivity) was 6 times greater along unpaved roads (17 times greater along paved roads) than it was in nearby undisturbed areas.

The greater vegetation cover typically observed along roadsides also is often due, in part, to greater species richness in those areas (Holzapfel and Schmidt, 1990); however, much of this diversity may be represented by non-native species easily dispersed along roads and trails (Wilcox, 1989; Tyser and Worley, 1992; Parendes and Jones, 2000). Furthermore, local-scale increases in species richness can be associated with decreases in species richness at the landscape scale, thus creating a relatively impoverished and anthropogenic vegetation community (Holzapfel and Schmidt, 1990). Interestingly, increased vegetation cover along roadsides may attract more invertebrates and other organisms. For example, Lightfoot and Whitford (1991) found that shrubs along a road supported greater numbers of foliage arthropods. What is not clear, however, is whether high densities of animals in roadside habitats represent improved conditions for native fauna or dominance by invasive and/or non-native organisms. Furthermore, high densities do not necessarily represent population sources (that is, where survivorship and productivity are high enough to contribute to the species' overall population); instead, high densities can indicate poor-quality habitat into which subordinate animals may crowd and experience poor survivorship if they cannot find or defend better habitat (population sinks). In other words, density can be a misleading indicator of habitat quality (Van Horne, 1983). It is important to note, however, that the greater vegetation cover along roadsides compared to plots away from roads may be a phenomenon found only in arid environments (Hillel and Tadmor, 1962; Holzapfel and Schmidt, 1990).

Fugitive dust raised by OHV traffic also affects vegetation in the vicinity of roads. Along Alaskan roads heavily traveled by various types of vehicles, Walker and Everett (1987) found significant dust impacts up to 10 m (10.9 yd) from the roadside and dust blankets up to 10 cm (3.9 in) thick on mosses and other vegetation of low stature. Several morphological factors contribute to plant susceptibility to heavy dust loads, including mat or prostrate growth form, lack of a protective stem cortex or leaf cuticle, and intricate branching or closely spaced leaves that tend to trap dust (Walker and Everett, 1987; Spellerberg and Morrison, 1998). Processes that may be affected by dust include photosynthesis, respiration, and transpiration due to blocked stomata and cell destruction (Spellerberg and Morrison, 1998), all of which could result in reduced plant growth, size, productivity, and/or survivorship.

2.3.4. Annotated Bibliography for OHV Effects on Vegetation

Adams, J.A., Stolzy, L.H., Endo, A.S., Rowlands, P. G., and Johnson, H.B., 1982, Desert soil compaction reduces annual plant cover: California Agriculture, v. 36, no. 9–10, p. 6–7.

Soil crust properties and associated changes in vegetation composition were measured under controlled conditions over two 6-month wet seasons to determine how rapidly they were altered by vehicle passes. Reductions in annual plant cover occurred in tracks created by as few as 1 (on wet loamy sand) to 20 vehicle passes (on dry loamy sand). This cover reduction, however, was not due to fewer plants; rather, the plants were smaller, and their size depended on the duration of drying periods, during which the soil strength intensified in impact/track areas. Cover of annuals with large taproots (for example, *Chaenactis fremontii*) decreased in vehicle tracks, whereas the cover of *Schismus barbatus*, a grass with a fibrous root system, was significantly greater in tracks generated by 1, 3, 10, and 20 vehicle passes. It was determined that root growth for plants stops at soil strengths of about 20 kg/cm^2 (284.4 lb/in^2 [or PSI]). Soil disturbance also increased the potential for grasses and pioneering annual species to become established, whereas perennial species would take at least 5 years to return. A possible reason for this may be greater water availability in the track.

Angold, P. G., 1997, The impact of a road upon adjacent heathland vegetation—Effects on plant species composition: Journal of Applied Ecology, v. 34, no. 2, p. 409–417.

The effect of a road on heathland vegetation was investigated at five sites adjacent to the main trunk road through the New Forest, Hampshire, United Kingdom, and nine supplementary sites adjacent to five minor roads. There was enhanced growth of vascular plants near the road, notably heather and grasses, which was probably due to nitrogen oxides from vehicle emissions. There was a decrease in the abundance and health of lichens near the road. There was an increase in the abundance of grasses in the heathland near roads, which may be due to the changes in relative competitive ability of plant species under conditions of eutrophication. The extent of the edge effect in the heath was closely correlated with traffic intensity, with a maximum edge effect of 200 m adjacent to a dual carriageway.

Benninger-Traux, M., Vankat, J.L., and Schaefer, R.L., 1992, Trail corridors as habitat and conduits for movement of plant species in Rocky Mountain National Park, Colorado, USA: Landscape Ecology, v. 6, no. 4, p. 269–278.

Ground-layer vegetation was sampled along selected trail corridors to determine whether corridors provide habitat for certain species and serve as conduits for species dispersal. Patterns of plant species composition were analyzed in relation to distance from trail edge, level of trail use, and distance from trailheads, junctions, and campgrounds. Species composition was significantly affected by distance from trail edge and level of trail use, as species were favored or inhibited by the corridor, depending upon their growth habits. Species composition also was affected by distance from trailheads. These findings, along with the presence of exotic species, indicate that trail corridors in Rocky Mountain National Park function as habitat and conduits for dispersal of plant species.

Bolling, J.D., and Walker, L.R., 2000, Plant and soil recovery along a series of abandoned desert roads: Journal of Arid Environments, v. 46, no. 11, p. 1–24.

To elucidate factors controlling desert succession, soil and vegetation dynamics were examined along roads abandoned for 5, 10, 21, 31, 55 and 88 years in southern Nevada. None of the measured soil or vegetation parameters varied significantly with road age. Differences were found, however, between soils and vegetation on roads compared to those on nearby control sites, and soils differed between roads created by surface vehicular traffic and those made by bulldozing. Studies of recovery following disturbance in deserts must take into account natural patterns of plant and soil heterogeneity and initial disturbance type.

Brooks, M.L., 1995, Benefits of protective fencing to plant and rodent communities of the western Mojave Desert, California: Environmental Management, v. 19, no. 1, p. 65–74.

This paper documents the response of plant and small mammal populations to fencing constructed to preclude OHV activities between 1978 and 1979 at the Desert Tortoise Research Natural Area, Kern County, California. Aboveground live annual plant biomass was generally greater inside than outside the fenced plots during April 1990, 1991, and 1992. The non-native grass, *Schismus barbatus*, was a notable exception, producing more biomass in the unprotected area. Forb biomass was greater than that of non-native annual grasses inside the fence during all 3 years of the study. Outside the fence, forb biomass was significantly greater than that of nonnative grasses only during spring 1992. Percent cover of perennial shrubs was greater inside the fence than outside, while no significant trend in density was detected. There was also more seed biomass inside the fence, which may have contributed to the greater species diversity and density of Merriam's kangaroo rats (*Dipodomys merriami*), long-tailed pocket mice (*Chaetodipus formosus*), and southern grasshopper mice (*Onychomys torridus*) in the protected area. These results show that protection from OHV disturbance has many benefits, including greater overall community biomass and diversity.

Holzapfel, C., and Schmidt, W., 1990, Roadside vegetation along transects in the Judean Desert: Israel Journal of Botany, v. 39, p. 263–270.

Vegetation was studied on comparable plots along roadsides and in the surrounding area. The uniqueness of roadside vegetation was shown using indices and measurements that allowed comparison along a climatic gradient. Near roads, biomass and species diversity were notably greater than in surrounding areas, and the chorological composition was different, at least under arid conditions. The reasons for these differences are discussed based on investigations of site conditions. Increased water runoff and more favorable soil conditions seem to have had important influences on the vegetation community.

Kutiel, P., Eden, E., and Zhevelev, Y., 2000, Effect of experimental trampling and off-road motorcycle traffic on soil and vegetation of stabilized coastal dunes, Israel: Environmental Conservation, v. 27, no. 1, p. 14–23.

The aim of this study was to assess the response of soil and annual plants of stabilized Mediterranean coastal dunes in Israel to various intensities of short-duration pedestrian and motorcycle traffic. Experimental procedures entailed 0, 20, 50, 100, and 200 straight and 150 turn motorcycle passes. The response of annual plants was assessed by measuring ground cover, height, and species richness and diversity, and soil response was assessed by measuring penetrable depth, organic matter, and moisture content. Motorcycle passage had an immediate significant impact on annual plants at all traffic intensities. The maximum effect on plants was observed in the wheel tracks and in the turn lanes. Mean annual ground cover and height were less sensitive measures than species richness and diversity for determining the overall impact of motorcycles on the area.

Prose, D.V., Metzger, S.K., and Wilshire, H.G., 1987, Effects of substrate disturbance on secondary plant succession—Mojave Desert, California: Journal of Applied Ecology, v. 24, no. 1, p. 305–313.

The effects of substrate disturbance on perennial plant succession in the Mojave Desert were assessed at three military camps abandoned for 40 years. Soil compaction, removal of the top layer of soil, and altered drainage channel density caused significant changes in perennial plant cover, density, and relative species composition. Long-lived species, predominantly *Larrea tridentata*, were dominant in all control areas, but percent cover and density were greatly reduced in areas where substrate alterations were significant. At one camp where substrate alterations were insignificant in disturbed areas, *Larrea* was the dominant species (as it was in the control areas).

Schultink, G., 1977, Impact analysis of off-road-vehicle use on vegetation in the Grand Mere Dune environment: East Lansing, Michigan, Michigan State University, Report no. NASACR155764, 10 p.

A linear regression of percent unvegetated land in OHV-impacted areas versus time for two sample areas indicated that the areas underwent average declines of 1.9 and 5.9 percent per year in vegetation cover. Two factors were assumed to play roles in the difference: the difference in accessibility and the extent of vegetation fragmented during the first year of the study (one sample area was located closer to potential access points and was more fragmented initially).

Wilshire, H.G., Shipley, Susan, and Nakata, J.K., 1978, Impacts of off-road vehicles on vegetation: Transactions of the North American Wildlife and Natural Resources Conference, v. 43, p. 131–139.

Observations of the impacts of off-road vehicles on soils and vegetation were made at more than 400 sites in seven western states during 3 years. This type of land use had both direct and indirect effects on vegetation. Direct effects included crushing and uprooting plants. Indirect effects included modification of the soil, which affected plants beyond the areas directly impacted by vehicles, and restoration of the plant cover was inhibited. This paper covers the erosional effects on vegetation, depositional effects on vegetation, and the effects of physical and chemical modification of remnant soils on revegetation.

2.4. OHV Effects on Wildlife and Habitats: Native, Threatened, and Endangered Species

2.4.1. Section Summary

The impacts of OHV activities on wildlife and their habitats are numerous and well documented. Networks of roads and trails fragment habitat, reduce patch size, and increase the ratio of edge to interior. This may have serious consequences for area-sensitive species (those that cannot carry out certain aspects of their life cycles without large blocks of habitat or corridors linking habitat patches), predator-prey relationships, and overall population dynamics. In particular, fragmentation and edges created by OHV routes may have strong effects on animal movement patterns. Precluding or inhibiting animal movements effectively diminishes dispersal to and recolonization in other areas, thus increasing the likelihood of local extirpations. Overall, studies demonstrate that even narrow roads (paved and unpaved) and trails can represent significant barriers to the movements of animals. Reluctance to cross even narrow trails similar in width to routes created by OHV travel may alter or preclude the movements of various species. The cumulative effects of OHV-route networks proliferating across the landscape may have serious ecological consequences for species reluctant to cross OHV routes. Where threatened and endangered species are at risk, understanding their particular responses to roads of varying types, widths, use intensities, and habitat contexts is crucial.

OHV routes also generate conditions unlikely to occur in environments unaffected by OHV activity; in turn, these conditions can facilitate range extensions and invasions of non-native and/or opportunistic species. In addition, OHVs can contribute directly to mortality (and possible population declines) of wildlife species through collisions with vehicles, nest destruction, and collapsing burrows. Noise generated by OHVs also has been found to cause inner ear bleeding. In particular, noise may alter animal behaviors, breeding populations, the abilities of some species to detect predators (through auditory cues), and it can stimulate estivating animals to emerge from their underground burrows at inappropriate times. These factors may result in diminished body mass, reduced productivity, and/or poor survivorship.

2.4.2. Loss of Habitat Connectivity: Fragmentation and Barrier Effects

Creating roads and trails (of any kind) diminishes habitat connectivity, increases the proportion of edge to interior habitat, and decreases patch size of habitats (Reed and others, 1996; Forman and others, 2003). In fact, roads, including OHV routes, represent a principal factor contributing to habitat fragmentation at various scales (Meffe and Carroll, 1997). Furthermore, both paved roads and OHV routes—ranging from 4-lane paved highways to two- track routes less than 3 m (3.3 yards) wide—that separate once-continuous habitat can disrupt the movement and dispersal of many wildlife species between and within habitats (Swihart and Slade, 1984; Brody and Pelton, 1989; Yanes and others, 1995; Lovallo and Anderson, 1996; Clevenger, 1998; Forman and Alexander, 1998; Jackson and Griffen, 1998). In turn, these effects can have consequences for area-sensitive species and may encourage non-native and/or invasive species. Special-status wildlife species known to occur on BLM lands and whose long-term persistence is threatened by habitat fragmentation and diminished habitat connectivity include grizzly bear (*Ursus arctos horribilis*; Gibeau and Herrero, 1998; Servheen and others, 1998), black bear (*Ursus americanus*; Brody and Pelton, 1989), gray wolf (*Canis lupus*; Paquet and Callahan, 1996), mountain lion (*Felis concolor*; Beier, 1993), lynx (*Felis lynx*; Ruediger, 1998), ocelot (*Leopardus pardalis*; Tewes and Blanton, 1998), and desert tortoise (*Gopherus agassizii;* Boarman and Sazaki, 1996). The resulting isolation of subpopulations (Dobson and others, 1999) can promote increased inbreeding and a lack of genetic exchange with other subpopulations, ultimately leading to declines in the genetic diversity required for adaptation to variable conditions and possible founder effects (Hanski and Simberloff, 1997; Hanski, 1999). Another consequence of subpopulation isolation is the reduced potential for recolonization when extirpations occur as a result of localized population fluctuations and catastrophic events (Yanes and others, 1995).

Until recently, only wide, multi-lane, paved roads have been considered significant barriers to animal movements. More recent lines of evidence from fragmentation studies, however, indicate that the ability or willingness of an animal to cross a given road type varies widely by species (Brody and Pelton, 1989; Lovallo and Anderson, 1996). For example, rodents in a desert habitat were found to avoid crossing a 4-lane highway, although they lived alongside the road in the right-of-way vegetation (Garland and Bradley, 1984). Likewise, in forested habitats divided highways wider than 90 m (98.4 yd) served as total barriers to dispersal by small forest mammals (Oxley and others, 1974). However, improved gravel roads have been found to inhibit crossings by mountain lions (*Puma concolor*; van Dyke and others, 1986), and even infrequently traveled, single-lane dirt roads have been found to alter movements by some species (Andrews, 1990). For example, Swihart and Slade (1984) report that prairie voles (*Microtus ochrogaster*) and cotton rats (*Sigmodon hispidus*) were strongly inhibited from crossing a route less than 3 m (3.3 yd)wide and composed of two dirt tracks created by the passing of 10 to 20 vehicles per day. Oxley and others (1974) evaluated small mammal responses to roads and routes ranging from 4-lane paved highways to country gravel roads in forested systems of southeastern Canada and found that they were not willing to cross roads or other routes with a total clearance (the distance between forest margins, including road surfaces and immediately adjacent strips of vegetation kept very short via spraying and/or mowing) of 30 m (32.8 yd) or greater; road surface apparently was unimportant. Likewise in Germany, forest mice (*Apodemus flavicollis*) did not cross roads 6 m (6.6 yd) wide, and very few mice returned to the side of the road from which they were captured after being translocated to the opposite side within the same habitat type (Mader, 1984). Areas characterized by high densities of roads also are characterized by low probabilities that amphibian species will occupy breeding pools (Vos and Chardon, 1998), most likely because the edges were relatively impermeable (whether due to behavioral avoidance or direct mortality) to critical amphibian movements (dispersal, seasonal movements; Gibbs, 1998). On the other hand, some small mammals are known to cross paved and gravel roads (Bakowski and Kozakiewicz, 1988), particularly where vegetated highway right-of-ways resemble those of adjacent habitats (Wilkins, 1982). These studies indicate that road surface type is not always the critical inhibiting factor; however, it does influence traffic speed, which can directly affect mortality rates (Oxley and others, 1974; Bakowski and Kozakiewicz, 1988).

Invertebrates also may be precluded from crossing various road types, including those considered relatively narrow; again, however, there are species differences that may be influenced by their ecologies and physical capabilities. For example, Samways (1989) found that both "tarred" (paved) and "untarred" roads were almost complete or partial barriers to three species of bush crickets (*Decticus varrucivorus monspeliensis* and *Platycleis fedtschenkoi azami*, both wingless, and *P. tessellate*, the flight range of which is less than [<] 5m [5.5 yd]), but roads were only minor, very minor, or did not serve as barriers to the movements of six other bush cricket species, five of which can readily fly across roads (flight ranges from <30 to 150 m [32.8–164.0 yd]. On the other hand, Munguira and Thomas (1992) found that wide highways did not affect the movements of butterflies in open populations; movements of butterflies in closed populations, however, were slightly impeded by roads.

Other butterfly species may not even attempt to fly across roads (described by authors as two-lane highways and secondary roads), possibly due to the extreme changes in microclimate over roads (including columns of warm air rising above roads; Boer Leffef, 1958, as interpreted and translated by van der Zande, 1980). Mader (1984) reported that in a five-year mark-recapture-release study involving 10,186 carabid beetles representing nine species, three species were never recaptured on the opposite side of study area roads (one- or two-lane paved roads) or parking loops, and the remainder were recaptured across the road only rarely. However, some individuals of a Swedish snail species (*Arianta arbustorum*) that were captured and translocated to the opposite sides of narrow paths or relatively wider roads did return to the capture sides of paths (Baur and Baur, 1990).

2.4.3. Edge Effects

Aside from fragmenting habitat, roads and trails of any kind also create habitat edges (Reed and others, 1996). In many instances, these edge effects extend well beyond the road's actual footprint and for some species the effects may extend well into the desert interior. Therefore, assessing edge effects of roads and trails on wildlife may entail determining distributions of wildlife in reference to the extent of any one edge effect (Yahner, 1988). Even then there may be an array of factors that vary the distances from roads/trails at which edge effects may be apparent. For example, Nicholson (1978) indicates that metapopulations of desert tortoises may be depleted within 0.8 km (0.5 mi) of highway edges, and von Seckendorff Hoff and Marlow (1997) indicate that this effect may extend as far as 3.5 km (2.2 mi) from the highway edge.

Given the frequent incidence of significant vegetation cover along road edges, many organisms may be attracted to right-of-way habitats. For example, Adams and Geis (1983) found greater small mammal density within interstate right-of-way habitats than in adjacent habitats. Density, however, can indicate habitats sinks to which animals retreat when more desirable habitats are occupied (Van Horne 1983). Alternatively, road edges may serve as ecological traps (Andrews, 1990) that are attractive and replete with necessary resources on the one hand, but impose unusually high mortality rates on the other hand. For example, birds may be attracted to lush roadside vegetation for breeding, nesting, or foraging (Clark and Karr, 1979), but they may be at great risk of mortality due to being hit by vehicles (Mumme and others, 2000). Similarly, avian eggs and nestlings can experience increased mortality due to high rates of predation (Yahner and others, 1989) in edge habitats. As mentioned in the section above, edge effects along roads can alter or preclude the seasonal movements of amphibians to their breeding pools (Gibbs, 1998; Vos and Chardon, 1998).

In the same ways that travel routes promote increased dispersal of non-native and invasive plant species, they also promote increased distributions of wildlife species otherwise unlikely to be common in a given area; in turn, this exerts additional competitive pressures on native species. Huey (1941) documented pocket gophers (*Thomomys umbrinus*) extending their ranges across the Mojave Desert via roads and canal systems. Although much of the surrounding desert landscape contained soils unsuitable for gophers, the attractive habitat (greater cover of vegetation resulting from increased moisture availability) along roadsides and canals facilitated the spread of these animals (Huey, 1941). An additional important edge effect associated with roads of many types is the presence of utility infrastructures, which can contribute to significantly altered predator-prey relationships along roads. For example, raven species (*Corvus* spp.) have increased their distribution throughout the Mojave Desert, primarily due to the fact that they can perch along utility structures to scan for carcasses on adjacent roads (paved and unpaved) (Knight and Kawashima, 1993), a significant concern in light of the fact that Berry and others (1986) reported ravens as being responsible for 68 and 75 percent of mortality among juvenile desert tortoises on two study plots.

2.4.4. OHV Disturbance and Noise

Vehicular traffic is also a source of noise and other stimuli that have the potential for disturbing wildlife along any type of road or trail (Singer, 1978; van der Zande, 1980; Brattstrom and Bondello, 1983; Bowles, 1995; Reijnen and others, 1995, 1996; Bowles, 1995; Kaseloo and Tyson, 2004). Veen (1973; as interpreted and translated by van der Zande, 1980) found that four shorebird species inhabiting open grassland areas were disturbed within 500–600 m of a "quiet rural road" and within 1600–1800 m of a "busy highway;" van der Zande (1980) reanalyzed Veen's data and yielded similar results for three of the four species, and went on to conclude that populations of these birds were diminished by as much as 60 percent over those distances. Forman and Alexander (1998) found that noise levels generally increase with traffic intensity, and Reijnen and others (1995, 1997) concluded that traffic noise can lead to significant reductions in breeding bird densities. Larger animals also exhibit responses to the intensity of traffic and traffic noise. Lyren (2001) found that coyotes changed their road-crossing periods in response to changes in traffic intensity throughout the day, and Singer (1978) reported that, in response to the shifting of truck gears, mountain goats ran away from a road edge when the truck was 1 km (0.6 mi) away from them, and they ran away from a lick that was 400 m (437.4 yd) from the road.

Noise emitted from certain types of OHVs can be as high as 110 decibels, which is near the threshold of human pain (Lovich and Bainbridge, 1999). Although sounds from OHV motors are not the loudest anthropogenic sounds, in wildlife habitats they are emitted more frequently than other high-intensity sounds (Brattstrom and Bondello, 1983), and the effect on animals can be significant. For example, sand lizards (*Uma scoparia*) and kangaroo rats (*Dipodomys deserti*) experienced hearing loss that lasted for weeks after being exposed to less than 10 minutes of dune buggy playback recordings played intermittently at lower decibel levels than the animals would have been exposed to in the actual presence of a dune buggy (Brattstrom and Bondello, 1983); subsequently, both species were unresponsive to recordings of predator sounds. In two other studies, kangaroo rats (*Dipodomys spectabilis*) experienced inner ear bleeding when subjected to OHV noise (Berry, 1 980b; Bury, 1980). Another issue is the way in which OHV noise (sound pressure) may simulate that of natural sounds (thunder, for example) to which many animals may be adapted to respond. For example, in response to 30 minutes of taped motorcycle sounds, Brattstrom and Bondello (1983) documented a spadefoot toad (*Scaphi opus couchii*) emerging prematurely (wrong season, absence of rain) from its burrow, most likely because the sound mimicked that of thunder, to which the species would normally respond.

Noise, lights, and other disturbances associated with OHV activities also have the potential for eliciting stress responses from a broad spectrum of wildlife taxa. Indeed, studies have shown that ungulates, birds, and reptiles all experience accelerated heart rates and metabolic function during disturbance events; in turn, animals may be displaced and experience reproductive failure and reduced survivorship (see review in Havlick, 2002). For example, radio- collared mule deer disturbed by ATVs altered their patterns of foraging and spatial use of habitat; deer in undisturbed areas, however, exhibited no such changes (Yarmoloy and others, 1988). In addition, Yarmoloy and others (1988) found that harassment of deer resulted in diminished reproductive output in the following fawning season, whereas deer that were not harassed experienced no change in reproduction.

2.4.5. Wildlife Mortality and Related Issues

Direct wildlife mortality can result from vehicular impact (Harris and Gallagher, 1989; Beier, 1993; Bruinderink and Hazebrook, 1996; Moore and Mangel, 1996), thus removing individuals from populations (Harris and Gallagher, 1989; Forman and Alexander, 1998); thus, habitats containing roads may represent population sinks for any species that commonly attempts to move from one habitat fragment to another by crossing roads (Kline and Swann, 1998). If mortality rates exceed rates of reproduction and immigration, wildlife populations decline (Beier, 1993; Bruinderink and Hazebrook, 1996; Moore and Mangel, 1996; Forman and Alexander, 1998). Previous studies indicate that mortality rates vary widely according to habitat and road or route characteristics (for example, road width, traffic density and speed, adjacent habitat) (Ward, 1982; Bashore and others, 1985; Foster and Humphrey, 1995; Evink and others, 1996, 1998), as well as taxa studied—invertebrates: Seibert and Conover (1991), Munguira and Thomas (1992); reptiles and amphibians: Rosen and Lowe (1994), Ashley and Robinson (1996), Boarman and others (1998), Rudolph and others (1998), Means (1999); birds: Dhindsa and others (1988), Moore and Mangel (1996), Mumme and others (2000); and mammals: Gilbert and Wooding (1996), Romin and Bissonette (1996), Lehnert and Bissonette (1997), Gunter and others (1998), Lyren (2001). Even where the frequency of wildlife mortality is relatively low most of the year, it may increase during certain seasons (Feldhammer and others, 1986; Bruinderink and Hazebrook, 1996) or when traffic frequency increases (McCaffery, 1973). Furthermore, population dynamics can be altered if low mortality rates nonetheless cause disproportionate mortality among specific sex and/or age classes (Beier, 1993; Moore and Mangel, 1996; Mumme and others, 2000).

Several researchers have conducted extensive monitoring at desert OHV sites and undisturbed sites to compare direct effects of OHV activity on mortality and abundance of certain reptile species (Bury and others, 1977; Berry, 1980a; Bury, 1980; Luckenbach and Bury, 1983; Brooks, 1999; Grant, 2005). Of important concern is the susceptibility of desert tortoises to mortality on all types of roads. Berry (1980a) found a link between OHV activity and population declines of the desert tortoise and Couch's spadefoot toad (*Scaphiopus couchii*); numbers of tortoises and active burrows in a 25-ha control plot were significantly greater than in a similar plot exposed to OHV activity, presumably the result of direct mortality from vehicles or the collapsing of burrows caused by OHV traffic (Lovich and Bainbridge, 1999). Additionally, the body masses of subadult and adult tortoises in the control plot were greater than those of tortoises in the OHV area (Bury and Luckenbach, 1986, cited *in* Lovich and Bainbridge, 1999). When comparing lizards in OHV-impacted plots to control plots, controls supported 1.8 times more species, 3.5 times more individuals, and 5.9 times more biomass (Luckenbach and Bury, 1983). Similarly, Bury and others (1977) found more reptile species (1.63 times more) and greater reptile abundance (182 percent more individuals) at control sites than at OHV sites. In another study, the remains of 39 tortoises were recorded during three surveys over a 2.5-year period along a 24-km (14.9-mi) section of paved highway in the western Mojave Desert (Boarman and others, 1993). Snakes also experience high rates of mortality in the Mojave Desert due to their strategy for thermoregulation (lying on warm surfaces, such as roads; Sullivan, 1981). Rosen and Lowe (1994), who conducted nighttime snake surveys along a 2-lane paved road in the Sonoran Desert (primarily within Organ Pipe Cactus National Monument), documented a 72 percent rate of snake mortality (104 live, 264 dead); mortality peaked in spring—when snake activity was moderately high and automobile traffic had not yet reached its summer minimum—and during rain events in the monsoon season (July through early September). Overall snake mortality during the entire 4-year study was estimated at 2,383 snakes (13.5 snakes/km/year; 8.1 snakes/mi/year), although actual numbers were likely closer to 4,000.

Densities and species diversity of desert birds and small mammals also have been reported to decrease in areas where OHV use was extensive (Busack and Bury, 1974; Bury and others, 1977; Luckenbach, 1978; Luckenbach and Bury, 1983; Brooks, 1999). Direct and indirect effects of OHVs on these species include breaking shrubs containing nests (nests, eggs, or nestlings destroyed) and diminished cover when shrubs are reduced or eliminated, mortality due to vehicle impact (especially ground-dwelling animals), and collapse of burrows due to OHV traffic (Bury and others, 1977). Bury and others (1977) found greater small mammal species richness (1.25 times greater) and abundance (500 percent more individuals) at control sites than OHV sites. Similarly, Luckenbach and Bury (1983) found 1.5 times more small mammal species, 5.1 times more individuals, and 2.2 times more biomass in control plots than in OHVimpacted plots; the number of desert kangaroo rats recorded in OHV plots was 53 percent lower than the number in control plots. Luckenbach and Bury (1983) found that overall animal activity—as measured by track frequencies—was greater in control areas than it was in OHV-use areas: arthropod tracks were 24 times more abundant, kangaroo rat tracks were 5 times more abundant, kit fox tracks were 2 times more abundant, and cottontail rabbit tracks were 10 times more abundant. Finally, Brooks (1999) found that protected areas in the Desert Tortoise Research Natural Area supported a greater abundance and species richness of birds and lizards than nearby portions of the desert subjected to intense OHV use and past sheep grazing. In one study, however, road mortality did not appear to have detrimental effects on densities of small mammals inhabiting highway right-of-ways, although the authors admit that they could not rule out confounding effects of immigration (Adams and Geis, 1983). In a study of 36 radio-marked flat-tailed horned lizards (*Phrynosome mcallii*) subjected to high (60 percent OHV track coverage in 60 minutes of riding time), low (30 percent in 20 minutes of riding time), and no (0 percent) impact by OHV traffic in 100 × 100 m (109.4 × 109.4 yd) plots, all survived. At the time of OHV treatment, however, 32 of the lizards were in their hibernation burrows 2–17 cm (0.8–6.7 in) underground (Grant, 2005), and it remains unclear whether soil substrates and vegetation growing above the burrows helped protect the animals from being crushed (21 of the 32 were under shrubs, and 8 burrows were known to have been run over directly by OHVs).

A major indirect effect of OHV activity on vertebrate survivorship is loss of vegetation cover. For all terrestrial vertebrates sampled, including species of conservation concern (desert kangaroo rat [*Dipodomys deserti*] and fringe-toed lizard [*Uma notata*]), Bury and others (1977) found a positive correlation between the percent canopy cover of creosote bush and species richness, abundance, and biomass. In a study of OHV effects on biota (including herbaceous and perennial plants, arthropods, lizards, and mammals) of the Algodones Dunes area in California, Luckenbach and Bury (1983) detected 9.4 times more cover, and 40 times more overall volume in control plots than in OHV-impacted plots, largely because shrubby perennial cover was greater in control plots. Another indirect effect of OHV activity on wildlife mortality is the proliferation of routes that provide greater access to remote places by hunters, poachers, and people seeking several forms of nonconsumptive recreation (Boyle and Samson, 1985; Andrews, 1990). Boyle and Samson (1985) also report a variety of nonconsumptive recreation impacts on wildlife, including flushing animals off nests; unnecessary energy expenditures; and displacement of animals from food, shelter, and other vital resources. Of particular concern was the increasing access that roads provide for tortoise collectors, which may explain declining trends in tortoise numbers along highways (Boarman and others, 1997).

2.4.6. Annotated Bibliography for OHV Effects on Wildlife and Habitats: Native, Threatened, and Endangered Species

Berry, K.H., 1980, A review of the effects of off-road vehicles on birds and other vertebrates, *in* DeGraaf, R.M., and Tilghman, N.G., eds., Management of western forests and grasslands for nongame birds—Workshop proceedings, Salt Lake City, Utah, February 11–14, 1980: Ogden Utah, U.S. Forest Service, Intermountain Forest and Range Experiment Station, General Technical Report INT–86, p. 451–467.

A review of the literature on the effects of off-road vehicles revealed that OHV use has significant effects and can reduce numbers, diversity, and biomass of birds and other vertebrates. The degree of impact depends upon amount and intensity of OHV use, habitat type, and sensitivity of the species.

Boarman, W.I., and Sazaki, M., 2006, A highway's road-effect zone for desert tortoises (*Gopherus agassizii*): Journal of Arid Environments, v. 65, no. 1, p. 94–101.

Roads can affect populations of animals directly (vehicle-animal collisions) and indirectly (due to habitat fragmentation and dispersal/proliferation of non-native or predatory species). This study investigated the effect of a 2- to 4-lane highway (with a posted speed limit of 65 mi/hr [105 km/hr] and an average daily traffic intensity of 8500 vehicles) on threatened desert tortoise (*Gopherus agassizii*) populations in the Mojave Desert, California, and attempted to determine the width of the road-effect zone by counting signs of tortoises (shells, tracks, scats, burrows, and pallets) along transects at 0, 400, 800, and 1600 m from and parallel to the edge of a highway. Mean sign count was 0.2/km (0.32/mi) at 0 m (0 yd), 4.2/km (6.72/mi) at 400 m (437.4 yd), 5.7/km (9.12/mi) at 800 m (874.9 yd), and 5.4/km (8.64/mi) at 1600 m (yd) from the highway edge. The differences between all distances except 800 and 1600 m (874.9 and 1749.8 yd) were statistically significant, suggesting that tortoise populations in the study area were depressed within a zone extending at least 400 m (437.4 yd) from the highway.

Brattstrom, B.H., and Bondeio, M. C., 1983, Effects of off-road vehicle noise on desert vertebrates, *in* Webb, R.H., and Wilshire, H.G., eds., Environmental effects of off-road vehicles—Impacts and management in arid regions: New York, Springer-Verlag, p. 167–206.

This study determined that sand lizards (*Uma scoparia*) and kangaroo rats (*Dipodomys deserti*) suffered hearing loss lasting for weeks after being exposed to less than 10 minutes of playback recordings of dune buggy sounds played intermittently at intensities lower than the average intensity levels actually emitted by OHVs. Such impacts led to the inability of both of these species to respond to recordings of predator sounds. A spadefoot toad (*Scaphi opus couchii*) emerged prematurely from its burrow when exposed to 30 minutes of taped motorcycle sounds.

Brooks, M.L., 1999, Effects of protective fencing on birds, lizards, and black-tailed hares in the western Mojave Desert: Environmental Management, v. 23, no. 3, p. 387–400.

Effects of a protective (fenced) area on birds, lizards, black-tailed hares (*Lepus californicus*), perennial plant cover, and structural diversity of perennial plants were evaluated from spring 1994 through winter 1995 at the Desert Tortoise Research Natural Area (DTNA), in the Mojave Desert, California. Abundance and species richness of birds were greater inside than outside the DTNA; these effects, however, were more pronounced during breeding season and a year of high rainfall than during winter and a year of low rainfall. Nesting activity was also more frequent inside the exclosure. Total abundance and species richness of lizards and individual abundances of western whiptail lizards (*Cnemidophorous tigris*) and desert spiny lizards (*Sceloporus magister*) were greater inside than outside the exclosure. Black-tailed hares generally prefer areas of low perennial plant cover, which may explain why they were more abundant outside than inside the DTNA. Habitat structure may not affect bird and lizard communities as much as availability of food at this desert site, and the greater abundance and species richness of vertebrates inside than outside the DTNA may correlate with abundances of seeds and invertebrate prey.

Bury, R.B., Luckenbach, R.A., and Busack, S.D., 1977, Effects of off-road vehicles on vertebrates in the California desert USA: Wildlife Research Report no. 8, U.S. Fish and Wildlife Service, Washington, D.C., p. 1–23.

This study compared differences in avian diversity, abundance, and biomass in unused and OHV-disturbed sites. Compared to OHV sites, reptile species richness was 1.63 times greater and there were 270 more individuals at control sites. Similarly, mammal species richness was 1.25 times greater and there were 115 more individuals at control sites than at OHV sites. The potential for ground nests of birds to be crushed and incubating birds to abandon nests was greater in areas of high OHV activity. Indirect effects of OHV activity on vertebrates were primarily caused by the loss of vegetation cover. There was a positive correlation between the cover of creosote bush and the total number of species, abundance, and biomass of all terrestrial vertebrates sampled.

Bury, R.B., and Luckenbach, R.A., 2002, Comparison of desert tortoise (*Gopherus agassizii*) populations in an unused and off-road vehicle area in the Mojave Desert: Chelonian Conservation and Biology, v. 4, no. 2, p. 457–463.

This study examined habitat, abundance, and life history features of desert tortoises (*Gopherus agassizii*) on two 25-ha plots in the western Mojave Desert: one unused and one used by OHVs. The unused plot had 1.7 times more live plants, 3.9 times more plant cover, 3.9 times more desert tortoises, and 4.0 times more active tortoise burrows than a nearby area used heavily by OHVs; these between-plot differences were all statistically significant. Furthermore, the few large-sized tortoises in the OHV plot had less body mass than those in the unused area. Although the scope of this study was limited to one paired-plot comparison, current data suggest that operation of OHVs in the western Mojave Desert results in major reductions in habitat and tortoise numbers, and possibly the body mass of surviving tortoises.

Lovich, J.E., and Bainbridge, D., 1999, Anthropogenic degradation of the southern California desert ecosystem and prospects for natural recovery and restoration: Environmental Management, v. 24, no. 3, p. 309–326.

Large areas of the southern California desert ecosystem have been affected by off-highway vehicle use, overgrazing by domestic livestock, agriculture, urbanization, construction of roads and utility corridors, air pollution, military training exercises, and other activities. Secondary contributions to degradation include the dispersal and proliferation of exotic plant species and a higher frequency of anthropogenic fire. Effects of these impacts include alteration or destruction of macro- and micro-vegetation elements, establishment of annual plant communities dominated by exotic species, destruction of soil stabilizers, soil compaction, and increased erosion. This paper provides a broad view of impacts on biota and cites several pertinent studies relative to OHV impacts on wildlife. The authors suggest that given the sensitivity of desert habitats to disturbance and the slow rate of natural recovery, the best management option is to limit the extent and intensity of impacts as much as possible.

Luckenbach, R.A., and Bury, R.B., 1983, Effects of off-road vehicles on the biota of the Algodones Dunes, Imperial County, California, USA: Journal of Applied Ecology, v. 20, no. 1, p. 265–286.

Algodones Dunes, the largest dune complex in California, contains many unique species; however, it also receives the greatest use by off-road vehicles in California. Studies of paired plots (unused versus OHV-impacted) and animal tracks along sand sweeps clearly demonstrated that OHV activities in the Algodones Dunes significantly reduced the biota. There were marked declines in herbaceous and perennial plants, arthropods, lizards, and mammals in OHV-used areas compared with nearby controls. All sand-adapted species, including several rare or threatened plants, were greatly reduced in habitats where OHVs operate; the biota was affected even by relatively low levels of OHV activity. Areas heavily used by OHVs had virtually no native plants or wildlife.

Rosen, P.C., and Lowe, C.H., 1994, Highway mortality of snakes in the Sonoran Desert of southern Arizona: Biological Conservation, v. 68, no. 22, p. 143–148.

A total of 368 snakes (104 live, 264 dead) were recorded over four years on a paved highway during 15,525 km (9,647.2 mi; mostly within Organ Pipe Cactus National Monument, Arizona) of driving along the road to detect amphibians and reptiles during rainfall events or while basking on the warm road surface. During 4 years, an estimated 2,383 snakes were killed on this stretch of pavement, although the actual number killed was probably closer to 4,000.

Webb, R.H., and Wilshire, H. G., 1983, Environmental effects of off-road vehicles—Impacts and management in arid regions: New York, Springer-Verlag, 534 p.

This book discusses the physical and biological effects of OHVs (recreational, mining, and military vehicles) on arid-land ecosystems, including effects on soils, vegetation, and wildlife. It also points out the loss of choices that OHV effects impose on future land users. Actual case studies are presented, complete with practical solutions, detailed planning measures that can be taken to reduce the adverse effects of OHVs, methods that can be used to rehabilitate the physical systems and vegetation communities of disturbed areas, and management concepts and practices that can be employed in protecting susceptible areas, including regulations and education.

2.5. OHV Effects on Water Quality

2.5.1. Section Summary

The direct effects of OHV activity on aquatic systems have received surprisingly little attention, due, in part, to the fact that OHV-impact research has focused on arid environments, where aquatic systems are seasonal or rare. Nonetheless, there is great potential for OHV activities to affect water quality in arid environs as well as well-watered regions. As described in Sections 2.2 and 2.3, soil properties and vegetation cover may be altered by OHV use; in turn, surface patterns of precipitation runoff (amount, velocity) may be altered, resulting in accelerated rates of erosion and sedimentation and elevated levels of turbidity in affected watersheds. Where slope is a factor, the extensive networks of OHV routes proliferating across landscapes can serve as conduits that direct or alter the direction of surface flows. These conduits may be eroded to form gullies that channel dislodged sediments and contaminants into aquatic ecosystems. Water quality also is adversely affected by OHV-raised dust that settles into aquatic systems.

OHV-dispersed chemicals also may be transported into aquatic systems. The operation of OHV engines, especially 2-stroke engines, can impact water quality through spills and emissions. These contaminants may enter aquatic systems via direct flushing, or they may be adsorbed to sediments and/or absorbed by plant materials, both of which are easily transported to aquatic systems by precipitation runoff or wind. Spill or emission contaminants may include 1,3- butadiene, benzene and ethylbenzene, xylenes, and toluene. Prior to the ban on leaded gasoline, lead levels were high in plants and animals near roads, and although the 1996 ban on leaded gasoline has resulted in dramatic declines in lead levels, it persists in the soil and may be mobilized when soils are eroded into wetlands.

2.5.2. Sedimentation and Turbidity

Areas naturally most susceptible to water-quality problems are those where infiltration rates are low, slopes are steep, the ratio of surface sand and gravel to finer particles is low, and where rainfall events are typically prolonged and intense (Iverson and others, 1981). Altering soil texture, disrupting soil crusts or desert pavement, and reducing vegetation cover can increase the soil's susceptibility to erosion; in turn, rates of sedimentation and turbidity levels can increase and alter the water quality of a given watershed, including streams and rivers, lakes, and small, isolated wetlands, including vernal pools (Forman and others, 2003). Sediments can displace the water-holding volume of a wetland, thus diminishing or eliminating the wetland's hydrological function (Luo and others, 1977). For example, where OHVs had traveled over the soil, Iverson and others (1981) found that surface runoff was 5 times greater and yielded 10-20 times more sediment than where soils were undisturbed.

Where OHV activity occurs, networks of OHV routes proliferate. Wheel cuts and tracks within these networks may serve as water conduits that channel and direct water flow containing sediments and contaminants into aquatic ecosystems (Wemple and others, 1996; Forman and others, 2003, p. 185–197). The generally impervious nature of soils compacted by OHV traffic enhances gully formation in these conduits, thus promoting additional flows of sediments and suspended solids into aquatic systems, effectively extending the drainage network of a given watershed, and potentially changing the timing of peak runoff flows (Wemple and others, 1996). The presence of OHV-route networks is an important factor in determining the severity of potential sedimentation in nearby aquatic systems. In particular, Wemple and others (1996) found that the drainage ditches along logging roads and the gullies that form below culvert outlets (where drainage flows pass under a road, or cross-drains) on steep slopes served as primary conduits linking surface flows to streams. The extent to which sediments might be carried along these conduits and into aquatic systems depends primarily on the presence of obstructions below cross drains and the spatial intervals between them (Haupt, 1959). In situations where cross drains were positioned at sufficient distances from streams, the drainage discharge infiltrated the soil and did not contribute to sedimentation in streams (Haupt, 1959). In areas characterized by soils with relatively low infiltration rates, such as those compacted by OHV use, transport of sediments over greater distances and into aquatic systems may be substantial.

Furniss and others (2000) describe similar effects of road and/or trail networks across a landscape. In particular, they discuss the continuous "hydrological connections" that facilitate sediment transport between surface flows and waterways. Furniss and others (2000) go on to list ways in which water and associated sediments enter stream systems from roads, including (1) inboard ditches (ditches perpendicular to the road footprint and that bisect the road) delivering runoff to a stream at a road-stream crossing, (2) inboard ditches delivering water to a cross-drain (culvert, dip, waterbar), (3) where sufficient discharge is available to create a gully or sediment plume that extends to the stream channel, (4) roads sufficiently close to streams so that the fillslope (road fill between the outside edge of the road and the base of the fill where it meets the natural ground surface) encroaches on the stream, and (5) landslide scars on the road fill. These connections provide direct routes for accelerated runoff transporting sediments and road- associated contaminants to natural drainage channels.

2.5.3. Dust and Contaminants

Water quality also is adversely affected when fugitive dust and contaminants enter aquatic systems. Emissions from OHVs, particularly those with 2-stroke engines, can include a variety of contaminants, which may settle directly in wetlands or they may be deposited in snow or directly on soils during rain events, from which they may be mobilized into wetlands.

Arnold and Koel (2006), who tested snowmelt runoff exposed to significant snowmobile emissions in Yellowstone National Park, detected benzene, ethylbenzene, m- and p-xylene, o-xylene, and toluene, and although all compounds were within the limits set by the U.S. Environmental Protection Agency, it is not clear what the cumulative impacts of these chemicals may be in watersheds. Adams (1975) found that the stamina of brook trout experimentally exposed to elements commonly found in snowmobile emissions, as measured by their ability to swim against the water current, was significantly diminished compared to that of control fish.

Airborne dust—and contaminants adsorbed to dust particles—raised by OHV traffic may eventually settle directly into wetlands (Forman and others, 2003, p. 231–234). The potential for adsorbed contaminants to be carried along with precipitation runoff and into wetlands is also a concern, as are plant materials containing absorbed contaminants. Finally, contaminants may enter aquatic habitats by direct flushing of exposed contaminants (for example, petroleum puddles). Prior to the ban on leaded gasoline, lead levels were high in plants and animals near roads (Daines and others, 1970; Motto and others, 1970; Quarles and others, 1974; Wheeler and Rolfe, 1979). Although the 1996 ban on leaded gasoline has since resulted in dramatic declines in lead levels along roadsides and in organisms, it persists in the soil and may be mobilized when soils are eroded into wetlands.

2.5.4. Annotated Bibliography for OHV Effects on Water Quality

Adams, E.S., 1975, Effects of lead and hydrocarbons from snowmobile exhaust on brook trout (*Salvelinus fontinalis*): Transactions of the American Fisheries Society, v. 104, no. 2, p. 363–373.

Prior to snowmobiling season, hydrocarbon levels in the water of a pond in Maine were undetectable; by the time of ice-out in spring, hydrocarbon levels had reached 10 parts per million (ppm) in the water and 1 ppm in exposed fish. In addition, exposed brook trout fingerlings contained 9 to 16 times more lead than control trout. Brook trout (*Salvelinus fontinalis*) held in aquaria for 3 weeks in melted snow containing three different concentrations of snowmobile exhaust also showed hydrocarbon and lead uptake. Stamina, as measured by the ability to swim against current, was significantly less in trout exposed to snowmobile exhaust than in control fish.

Brabec, E., Schulte, S., and Richards, P.L., 2002, Impervious surfaces and water quality—A review of current literature and its implications for watershed planning: Journal of Planning Literature, v. 16, no. 4, p. 499–514.

This paper focuses on the effect of impervious surfaces on the health of nearby aquatic habitats. Although considerable research has been done to define watershed thresholds of impervious surfaces (beyond which water quality declines), there are numerous flaws in the assumptions and methodologies used. Given refinement of the methodologies, accurate and usable parameters for preventive watershed planning can be developed, including thresholds of impervious surfaces and balances between pervious/impervious surfaces within a watershed.

Brown, K.J., 1994, River-bed sedimentation caused by off-road vehicles at river fords in the Victorian Highlands, Australia: Water Resources Bulletin, v. 30, no. 2, p. 239–250.

This study investigated some of the effects occurring at OHV crossings on two rivers in eastern Australia, where many road crossings occur at low-level fords. It provides a method whereby the amount of sediment redeposited downstream of a ford can be measured. Attention is drawn to the fact that sediment is contributed to rivers by five major processes: the exposure of surfaces, the concentration of surface runoff in wheel ruts, soil compaction and subsequent reduction of water infiltration leading to increased surface runoff, backwash from the vehicle, and undercutting of banks by bow-wave action. The last two of these processes have not been reported previously. Sediment collection experiments in two upland rivers indicated a mean deposition rate at the stream bed of approximately 1,000 g/m^2 over a period of 30 days.

Furniss, M.J., Flanagan, S.A., and McFadin, B.A., 2000, Hydrologically connected roads—An indicator of the influence of roads on chronic sedimentation, surface water hydrology, and exposure to toxic chemicals: U.S. Forest Service, Stream Systems Technology Center, Rocky Mountain Research Station, Technical Report, 4 p., *http://www.stream.fs.fed.us/streamnt/jul100/jul100_2.htm*.

This study defines the concept of forest-road drainage as a transport system for sediment into streams and proposes design changes to road drainage that would prevent or minimize this movement. The proportion of road that is hydrologically connected to a stream network may be a useful indicator of the potential for several adverse effects, including (1) the delivery of road-derived sediments to streams; (2) hydrologic changes associated with subsurface flow interception, concentration, and diversion; (3) increased drainage density; (4) extension of the stream network; and (5) the potential for road-associated spills and chemicals to enter streams.

Hamilton, L.J., 2002, A study of the effects of ORV stream crossings on water quality of two streams located in the Angelina National Forest, Texas—A physicochemical and benthic macroinvertebrate analysis: Masters Abstracts International, v. 40, no. 3, p. 668.

A study was conducted for the U.S. Forest Service to determine whether OHV-based stream crossings affected water quality of two streams located in Texas. The sites differed most in turbidity, total solids, Shannon's diversity index, dissolved oxygen, nitrate, and ratios of Chironomidae:EPT (Ephemeroptera + Plecoptera + Trichoptera, a common indicator of taxonomic richness detected during stream surveys to assess water quality), although there were no significant differences in the physicochemical properties. At one site, however, the upstream and downstream plots differed significantly in terms of two benthic indices—Hilsenhoff's *Biotic Index* and *Ratio of Scrapers to Filtering Collectors*.

Katz, M., Legore, R.S., Weitkamp, D., Cummins, J.M., and Anderson, D., 1972, Effects on freshwater fish: Journal of the Water Pollution Control Federation, v. 44, no. 6, p. 1226–1250.

This is a literature review of the effects of water pollutants on freshwater fish. Topics include (1) tests to determine the lethality of estuarine and some polluted river waters to trout and cyprinids; (2) estimated degrees of river pollution based on bacterial and chemical analysis of water samples; (3) documentation of some effects of municipal wastewater effluents on the water quality, fish populations, and bottom-fauna characteristics of a receiving stream; and (4) observations of the environmental effects of pollutants such as synthetic detergents, industrial wastes, and pesticides.

Roy, A.H., Rosemond, A.D., Leigh, D.S., Paul, M.J., and Wallace, J.B., 2003, Habitat-specific responses of stream insects to land cover disturbance—Biological consequences and monitoring implications: Journal of the North American Benthological Society, v. 22, no. 2, p. 292–307.

This study analyzed the impact of a range of physical and chemical stressors on aquatic insects and tested whether the effects of these stressors differed in three habitat types: riffles, pools, and banks. Riffle assemblages were affected by both physical (for example, streambed mobility) and chemical (specific conductance, nutrient concentration) variables. The density of aquatic insects in pools also was correlated to physical and chemical variables, but there were few relationships with pool or bank richness or bank density. Because relative impacts of disturbance in riffles were greater than in banks, the authors found greater differences between riffle and bank richness in streams with greater sedimentation. The proportion of bank richness (bank richness/bank + riffle richness) increased with finer bed sediment and increased bed mobility. The study also compared richness of facultative taxa (found in multiple habitats) between sites characterized as minimally impacted and sediment-impacted. In riffles, richness of facultative taxa was lower in sediment-impacted than in minimally impacted sites, but was similar for both disturbance groups in banks.

Wheeler, A.P., Angermeier, P.L., and Rosenberger, A.E., 2005, Impacts of new highways and subsequent landscape urbanization on stream habitat and biota: Reviews in Fisheries Science, v. 13, no. 3, p. 141–164.

This paper emphasizes a more thorough consideration of highway impacts and, ultimately, better land-use decisions by conceptualizing road development in three stages: initial construction, road presence, and eventual landscape urbanization. Road construction is characterized by localized physical disturbances, which generally subside through time. In contrast, road presence and landscape urbanization are characterized by persistent physical and chemical impacts. Though not specific to OHV activity, this paper does focus on the fact that landscape urbanization is clearly the greatest threat to stream habitat and biota, as stream ecosystems are sensitive to even low levels (less than 10 percent of a given watershed) of urban development. Researchers know little about the occurrence, loading rates, and biotic responses to specific contaminants in runoff from roads. Also needed is a detailed understanding of how drainage crossings, especially culverts, affect fish populations via constraints on movement and how road networks alter natural regimes (streamflow, temperature).

2.6. OHV Effects on Air Quality

2.6.1. Section Summary

Fugitive dust raised by OHV traffic on unpaved roads/trails can contribute significantly to air-quality problems. Also problematic are OHV emissions, particularly from 2-stroke engines. Currently, many OHVs in use, including off-highway motorbikes and ATVs, run on 2- stroke engines, which do not burn fuel completely and produce significant amounts of airborne contaminants, including nitrogen oxides, carbon monoxide, ozone, aldehydes, and extremely persistent polycyclic aromatic hydrocarbons (PAH), including the suspected human carcinogen, methyl tert-butyl ether (MTBE). Some airborne contaminants settle onto plants or into soils and function as fertilizers, thus causing changes in plant community composition and altering growth rates. The accumulation of emissions contaminants is evident in the tissues of plants and animals exposed to them. Prior to the ban on leaded gasoline, lead also was prevalent in plants and animals near paved roads and other travel routes, and because it persists in the environment, it can still have impacts when contaminated soils are mobilized.

2.6.2. Fugitive Dust Raised by OHV Traffic

Fugitive dust (largely composed of lightweight soil particles, including silt and clay) suspended in the air may impact more total area than any other impact of roads (paved or unpaved; Forman and others, 2003), and it can have significant effects on ecosystems (Westec, 1979). Dust is created and raised into the air as OHVs disturb soil crusts, abrade and pulverize soils, and generate wind currents. Once soil surfaces are disturbed, wind erosion may increase the amount of debris flow (Lovich and Bainbridge, 1999). In 1973, satellite photos detected six dust plumes in the Mojave Desert covering more than 1,700 km^2 (656.2 mi^2); the plumes were attributed to destabilization of soil surfaces resulting from OHV activities (Nakata and others, 1976; Gill, 1996). Along roads in Alaska heavily traveled by various types of vehicles, Walker and Everett (1987) found that dust had buried mosses and very low-statured vegetation in the 10- m-wide area adjacent to each side of the road; dust blankets measured up to 10 cm (3.9 in) deep. Accumulations of dust on vegetation can disrupt photosynthetic and respiration processes, leading to reduced plant growth, reproduction, and survivorship.

2.6.3. Contaminants Associated with OHV Use

Before emissions controls on automobiles became significantly more effective, there was little concern about emissions from small engines; today, however, their relative contribution to air-quality problems is significant (see *http://www.egr.msu.edu/erl/Small%20Engine%20Emissions.html)*. This is because small engines, especially 2-stroke models (many of which are being phased out), do not burn fuels completely; thus, their emissions contain the resulting by-products of incomplete combustion, including nitrogen oxides NO_x, sulfur dioxide (SO2), carbon monoxide (CO), ozone (O3), aldehydes, and extremely persistent polycyclic aromatic hydrocarbons (PAH). In fact, a very small, 2-stroke engine running for 2 hours emits the same amount of hydrocarbons as driving 10 cars (of the fuel-burning efficiency produced in 1995) for 250 miles each (*http://www. arb. ca.gov/msprog/offroad/sm_en_fs.pdf*).

Pollutants emitted from exhaust can cause a variety of impacts on vegetation. Carbon dioxide may function as a fertilizer and cause changes in plant species composition (Bazzaz and Garbutt, 1988; Hunt and others, 1991; Ferris and Taylor, 1995); nitrogen oxides also may function as fertilizers, producing similar effects along roadsides (Falkengren-Grerup, 1986; Holzapfel and Schmidt, 1990; Angold, 1997). Spencer and Port (1988) found that the soluble nitrogen content of perennial ryegrass (*Lolium perenne*) plants growing within 0-6 m (0-6.6 yd) of a paved road than in plants growing more than 6 m (6.6 yd) from the road, which contributed to greater growth rates and fecundity of aphids (*Rhopalosiphum padi*) inhabiting the plants closest to the road. Sulfur dioxide, which can be taken up by vegetation, may result in altered photosynthetic processes (Winner and Atkison, 1986; Mooney and others, 1988).

Several species of Mojave Desert perennials and annuals were fumigated in experimental chambers to determine their sensitivities to SO2, nitrogen dioxide (NO2), and O3; Thompson and others, 1980; Thompson and others, 1984). Creosote bush (*Larrea* sp.), the only perennial species found to be sensitive to SO2 and NO2, exhibited leaf injury and reduced growth when exposed to SO2 and NO2; however, numerous annuals, including redstem stork's bill (*Erodium cicutarium*) and desert Indianwheat (*Plantago insularis*; extremely sensitive), cleftleaf wildheliotrope (*Phacelia crenulata*; very sensitive), and wooley desert marigold (*Baileya pleniradiata*), exhibited more dramatic effects, including extensive injury and death (Thompson and others, 1980). Another study by Thompson and others (1984) revealed several annual species that are extremely sensitive to SO2 and O3, including brown-eyed primrose (*Camissonia claviformes*), Santa Cruz Island suncup (*C. hirtella*), and Nevada cryptantha (*Cryptantha nevadensis*).

OHV emissions also contain a variety of heavy metals, including zinc, copper, nickel, chromium, and lead (National Research Council, 1986). In terms of overall quantity, lead was one of the most significant heavy metals emitted prior to the ban on leaded gasoline in 1996 (Daines and others, 1970; Motto and others, 1970; Quarles and others, 1974; Wheeler and Rolfe, 1979). At least in desert regions, concentrations of lead particulates along roads were positively correlated with traffic volume (Motto and others, 1970). Within 80 m (87.5 yd) of roadsides, Quarles and others (1974) found that lead concentrations diminished notably from road edges (543 and 190 ppm) to 10 m (47 and 5 ppm, respectively) away from the edge; beyond 80 m, accumulations of lead diminished at lower rates. The declining gradient in lead concentrations away from roadsides may have been due, in part, to the direction of surface water flow (Byrd and others, 1983) as soil and other debris to which lead adheres were flushed away by the volume of water that runs off road surfaces. Although lead emissions from gasoline have declined dramatically since control policies were implemented in the 1970s (Forman and others, 2003), it persists in soils and can continue to move through the environment when contaminated soils are dislodged.

2.6.4. Annotated Bibliography for OHV Effects on Air Quality

Agrawal, Y.K., Patel, M.P., and Merh, S.S., 1981, Lead in soils and plants—Its relationship to traffic volume and proximity to highway (Lalbag, Baroda City): International Journal of Environmental Studies, v. 16, no. 3–4, p. 222–224.

Accumulations of lead from motor-vehicle exhausts on soils and trees growing along a busy thoroughfare in the Lalbag area of Baroda City were studied. Analysis of soils and tree samples showed that the distribution of emitted lead was influenced by the direction of the prevailing wind. Lead concentrations in plants and soils near the roadside were greater than they were in soils and plants 4–6 m away from the roadside.

Bazzaz, F.A., and Garbutt, K., 1988, The response of annuals in competitive neighborhoods— Effects of elevated CO2: Ecology, v. 69, no. 4, p. 937–946.

Four members of an annual plant community were used to investigate the effects of changing neighborhood complexity and increased carbon dioxide (CO2) concentration on competitive outcome. Plants were grown in monoculture and in all possible combinations of two, three, and four species in CO2-controlled growth chambers at CO2 concentrations of 350, 500, and 700 microliters/liter (μL/L) (1 ppm), with ample moisture and light. Species responded differently to enhanced CO2 level. The biomass of some species (*Abutilon theophrasti*, for example) increased with increasing CO2, while that of others (*Amaranthus retroflexus*) decreased with increasing CO2 concentration. The potential effects of CO2 on community structure could be profound, particularly at the intermediate levels of CO2 that are predicted for the first half of the 21st century.

Gish, C.D., and Christensen, R.E., 1973, Cadmium, nickel, lead, and zinc in earthworms from roadside soil: Environmental Science and Technology, v. 7, p. 1060–1062.

Cadmium (Cd), nickel (Ni), lead (Pb), and zinc (Zn) in soils and earthworms along two Maryland highways decreased with increasing distance (10, 20, 40, 80, and 160 ft) from the road. Along each highway, metal residues were greater where traffic volume was greater. Correlations between residues in earthworms and soil decreased with decreasing atomic weights (Pb, Cd, Zn, Ni). Metal residues in soils were positively correlated with quantities of soil organic matter. Earthworms accumulated up to 331.4 ppm of Pb and 670.0 ppm of Zn, concentrations that may be lethal to earthworm-eating animals.

Motto, I.L., Daines, R.I., Chilko, D.M., and Motto, C.K., 1970, Lead in soils and plants—Its relationship to traffic volume and proximity to highways: Environmental Science and Technology, v. 4, p. 23 1–237.

Lead concentrations increased with traffic volume and decreased with distance from highways. Much of the lead was present as removable surface contamination on plants, and major effects were limited to the soil surface within 100 ft (30.48 m) of the highway.

Nakata, J.K., Wilshire, I.G., and Barnes, G.C., 1976, Origin of Mojave Desert dust plumes photographed from space: Geology, v. 4, p. 644–648.

OHV-raised dust has been an enormous problem in the Mojave Desert, as illustrated by satellite photos that revealed six dust plumes covering more than 1,700 km^2 (656.4 mi^2) of the western Mojave region in January 1973; the dust plumes were attributed to destabilization of ground surfaces, primarily from OHV activity.

Quarles, H.D., Hanawalt, R.B., and Odum, W.E., 1974, Lead in small mammals, plants, and soil at varying distances from a highway: Journal of Applied Ecology, v. 11, no. 3, p. 937–949.

Lead particulates were measured at varying distances from three highways. Lead concentrations were greatest within 10 m (10.9 yd) of the highways. Lead concentrations in the soil along two transects dropped from 543 ppm and 190 ppm at the road edge to 47 ppm and 5 ppm 10 m from the road edge. Both plants and animals were susceptible to lead uptake.

Spencer, H.J., and Port, G.R., 1988, Effects of roadside conditions on plants and insects. II. Soil conditions: Journal of Applied Ecology, v. 25, no. 22, p. 709–7 15.

An experiment was done to investigate the performance of plants (*Lolium perenne*) grown in roadside soil. Significantly fewer plants germinated in soil taken 0 to 6 m from the road compared with soil taken 6 m from the road. For a given population size, however, plants grown in soil taken from beside the road attained significantly greater dry weight and significantly greater soluble nitrogen content. Nitrogen oxide emissions, identified as the probable cause of these effects, were absorbed by the roadside soil and subsequently assimilated by the plants.

Thompson, C.R., Olszyk, D.M., Kats, G., Bytnerowicz, A., Dawson, P.J., and Wolf, J. W., 1984, Effects of ozone or sulfur dioxide on annual plants of the Mojave Desert: Journal of the Air Pollution Control Association, v. 34, no. 10, p. 1017–1022.

Forty-seven species of annual plants from the Mojave Desert were grown in pots and exposed in open-top field chambers located at Riverside, California, to test their relative sensitivity to SO_2 and O_3. Species differed widely in their response to the pollutants. Three species, *Camissonia claviformis*, *Camissonia hirtella*, and *Cryptantha nevadensis*, were quite sensitive to both pollutants, exhibiting leaf injury when exposed to 0.1 ppm O_3 or 0.2 ppm SO_2. The other species were intermediate in sensitivity, and O_3 sensitivity did not always correspond to SO_2 sensitivity. For 8 of 11 species tested, total sulfur concentration was greater in plants exposed to 0.2 ppm SO_2 than in unexposed plants. *Baileya pleniradiata* and *Perityle emoryi* exhibited the greatest increases in sulfur concentration for exposed versus control plants.

Walker, D.A., and Everett, K.R., 1987, Road dust and its environmental impact on Alaskan taiga and tundra: Arctic and Alpine Research, v. 19, no. 4, p. 479–489.

The physical and chemical characteristics and ecological consequences of road dust in arctic regions were reviewed with emphasis on recent information gathered along the Dalton Highway and the Prudhoe Bay Spine in northern Alaska. Enhanced dust-control measures were considered, particularly where the road passes through scenic lichen woodlands, acidophilic tundra, and in calm valleys where dust commonly was a traffic-safety hazard.

Westec Services Inc., 1979, Fugitive dust impacts during off-road vehicle (ORV) events in the California desert: Tustin, California, WESTEC Services, Inc., Technical Report, 40 p.

Results and analysis of dust monitoring for five desert races demonstrated that factors such as distance from the point of generation, soil moisture, soil characteristics, wind, and relative humidity, as well as the type and number of vehicles in the race, had the largest effect on the amount and type of dust, particulate size, how quickly it settled, and the extent of the human health hazard present during the race. Long period (daily) dust-exposure levels were 10 times greater than the standard, whereas short period (hourly) dust levels were 100 times greater than the standard under adverse conditions near the race activity.

2.7. Socioeconomic Implications of OHV Use

2.7.1. Section Summary

The socioeconomics of OHV use include OHV user demands, concerns, and attitudes; the economic effects of OHV use on communities near OHV-use areas; the economics of managing OHV activities; the effects of OHV use on non-motorized recreators; and the economics of losing ecosystem services. Although not currently addressed through BLM's indicators of rangeland health, natural resource attributes are heavily influenced by socioeconomic factors. Since the mid 1980s, the incidence of OHV use on public lands has increased substantially, and this trend is expected to continue. Moreover, the economic benefits from travel expenditures and the sales of supplies and equipment in communities bordering OHV-use areas generates significant pressure to maintain or increase current levels of OHV activity. As OHV activity increases, however, increasing stress is placed on natural resources, land managers who must monitor and regulate OHV activities, and visitors seeking non-motorized forms of recreation.

2.7.2. Trends in OHV Use and Technology

In a survey of Utah OHV users commissioned by the Utah Department of Natural Resources, Fisher and others (2001) found that public lands are primary destinations among most users; only one quarter of survey respondents took trips to private land. More specifically, BLM land was the primary destination for ATV, motorcycle, and 4 x 4 vehicle users; U.S. Forest Service land was the secondary destination among ATV and 4 x 4 users; and State land was the secondary destination among motorcycle users (Fisher and others, 2001). Increasing OHV use is likely to be accompanied by greater demand for places where OHVs can be used, particularly near urban areas and corridors; as urban populations increase, so do the numbers of recreators on nearby public lands, thereby putting more stress on the landscape (Brooks and Champ, 2006). The increasing demands also pose problems for land managers already balancing the needs of a dynamic land base, often with limited budgets and/or staffing (Brooks and Champ, 2006; Rocky Mountain Research Institute, 2002). These limitations constrain land managers but not OHV use; thus, OHV recreation is largely "unmanaged." In addition, technology advancements in outdoor recreation equipment have led to production of OHVs that easily access lands previously unimpacted by mechanized recreation (Meine, 1998; Ewert and Shultis, 1999). As a result, new problems have arisen for both previously unimpacted areas and backcountry users who now encounter OHVs. Problems potentially arising from a constrained ability to manage lands include resource degradation, displacement of wildlife, and conflict among users, both within and across user types.

2.7.3. Types, Sources, and Effects of OHV User Conflict

Much of the OHV literature addresses conflicts between OHV users and other land users, even those who are not directly affected by OHV users. Researchers have addressed conflict issues by using a variety of tools or models designed to help managers understand and reduce conflicts between or among user groups. Bury and others (1983, p. 401) describe conflict as existing "whenever incompatible activities occur" and offer three elements that contribute to the incompatibility of activities: spatial and temporal proximity, dominance over the environment, and dependence on technology. When the proximity of activities does not result in direct or indirect (seeing the effects of other uses) encounters among user types, then environmental dominance and technological dependence are more likely to come into play. Dominance over the environment refers to how much an individual feels the need to exert some kind of control over the environment. Dependence on technology can cause conflict when people who retreat to backcountry to seek solace from modern technology clash with those who use technology to enhance their outdoor experiences. Conflict also occurs between land users and land managers. Inconsistent management policies across different land management agencies can cause such conflict, particularly as OHV recreation is ushered from being "unmanaged" to "managed." On many public lands, trails are currently considered open unless posted as closed, and once a trail has been established by users, it is often considered open for use (Brooks and Champ, 2006).

Graefe and Thapa (2004) outline some of the traditional approaches to examining user conflicts through research, including studies of goal interference (first introduced by Jacob and Shreyer, 1980). Goal interference occurs when a user comes into direct (seeing the conflicting recreation type) or indirect (seeing the *effects* of a recreation type) contact with another user type and is impeded from accomplishing the desired purpose of his or her recreation (Badaracco, 1976). The factors that contribute to goal interference are activity style, resource specificity, mode of experience (whether individuals are focused or unfocused), and tolerance for lifestyle diversity. Another model classifies conflict as either interpersonal conflict or a conflict of social values (Vaske and others, 1995). Interpersonal conflict is similar to goal interference in that a user has a problem with another use type and encounters an individual participating in, or evidence of, that type (hearing OHV noise, for example). Social values conflict occurs regardless of whether or not differing user types encounter one another—just knowing that the other recreation type is permitted may be unacceptable.

In the literature on user conflict, conflict is more often characterized as one-sided than two-sided (Badaracco, 1976; Bury and others, 1983; Watson and others, 1997; Graefe and Thapa, 2004). For example, while backpackers may perceive OHV users as disruptive to their experience, it is less likely that OHV users will find backpackers disruptive to their experience (Jackson and Wong, 1982). Displacement is the most common personal coping mechanism by which conflict is abated (Watson et al, 1997; Graefe and Thapa, 2004). That is, if an individual feels negatively enough about certain recreational activities occurring in the area he/she wishes to use, there is a possibility that the individual will simply forgo recreating in the area altogether, thereby increasing the probability that area managers will gradually lose support from that user base (Watson and others, 1997; Graefe and Thapa, 2004).

2.7.4. OHV Users and their Preferences

Overall, understanding the social effects of OHV use requires understanding the full array of recreational activities sought and the preferences of both OHV and non-OHV users alike. For example, people engaged in camping may include both OHV and non-OHV users, which can result in dissatisfaction among campers. In a survey of campers that included both OHV and non-OHV users, 66 percent indicated that having a regulated OHV riding area nearby would make their stay more enjoyable because it would reduce the number of riders in other areas and maintain a safer environment for both riders and campers (Bury and Fillmore, 1974). When given a choice between having (1) no motorcycle riding area but permission to ride on campground roads, (2) prohibition of all motorcycle riding, or (3) a nearby motorcycle area and no permission to ride on campground roads, 75 percent of riders and campers surveyed preferred the third alternative (Bury and Fillmore, 1974; riders and campers were socioeconomically similar).

Fisher and others (2001) reported that although 63.2 percent of motorcycle users surveyed did not stop to engage in any other type of recreational activity, almost 60 percent of ATV owners and 75 percent of 4 X 4 vehicle owners did engage in other recreational activities during their trips. Of those OHV users who did stop to engage in additional recreational activities, hiking was the most popular (>75 percent of motorcycle/4 X 4 vehicle users and 20 percent of ATV users). Hunting was the other most common recreational activity among ATV users and the second most common activity among 4 X 4 vehicle users; other recreational activities included fishing, camping, and sightseeing (Fisher and others, 2001).

Overall, the results of the user preference surveys discussed previously reveal a potentially conflicted OHV user base in that the quality of their associated recreational activities could be affected by OHV activities. For example, campers who wish to ride OHVs for additional recreation, but who feel strongly that OHV use should be restricted to designated areas, are likely to feel dissatisfied if other OHV users ride through the campground and/or on hiking trails. Similarly, if OHV use in preferred hunting or fishing areas—or other areas crucial to healthy populations of game and fish species—degrades habitat quality that results in diminished game and fish populations, then OHV riders who also hunt and fish may experience dissatisfaction.

Understanding the social effects of OHV activities (and potential outcomes of OHV activities) also requires determining where OHV users like to go and what their preferences are while riding. For example, in Colorado (where user attitudes are likely relatively moderate), Crimmins (1999) reported that

- 38.5 percent of OHV riders use U.S. National Forest Service land,
- 22.4 percent use private land,

- 18.6 percent use BLM land,
- 6.0 percent use State land,
- 3.4 percent use City or county land, and
- 2.3 percent use National Recreation Areas.

These data indicate that the use of public lands for OHV riding far outweighs that of private lands. Crimmins (1999) further reported OHV user preferences in terms of riding area attributes, which included

- no fee for use (if on public land),
- signs indicating all activities allowed on the trail, and
- locations removed from other human activity.
- The least important attributes included
- patrolling by staff of land management agencies or local OHV clubs,
- restrooms, and
- loading ramps (Crimmins, 1999).
- When presented with a list of priorities for uses of public funds, OHV users selected
- purchasing right-of-ways for OHV access,
- new OHV trail construction,
- erosion control, and
- OHV trail system planning and maintenance.

The low ranking of management patrols probably indicates that users desire more flexibility regarding where they may ride (Crimmins, 1999). Crimmins (1999) also pointed out that although the availability of facilities ranked low in terms of user preferences, management agencies nonetheless generally focus on providing facilities and generally report high user demand.

In terms of OHV user preferences for trail types and features, Bury and Fillmore (1974) reported that variation in terrain was the most important factor. The authors' recommendations for an effective OHV area included

- riding areas established near some, but not all, campgrounds;
- trails kept ≥600 feet (183 m) from the nearest campground;
- trails ≤6 feet (1.83 m) wide; and
- trails that traverse hillsides and include a variety of technical (obstacles, rugged terrain) and non-technical (no obstacles, smooth riding surface) features.

Fisher and others (2001) reported that motorcycle and ATV riders in Utah preferred

- riding off established trails (38.1 and 49.4 percent, respectively),
- double-track trails (12.7 and 17.1 percent),
- single-track trails (12.7 and 4.3 percent),
- moto-cross or ATV courses (9.5 and 15.1 percent), and
- roads (11.1 and 4.3 percent).

The issue of traveling off established trails is a serious concern with respect to natural resource management (Forman and others, 2003; Petersen, 2006); however, areas closed to OHVs and a shortage of designated OHV areas are common complaints among users (Achana, 2005; Fisher and others, 2001; Nelson and others, 2000). For example, in a survey commissioned by the Utah Department of Natural Resources to identify the most important issues affecting OHV use in Utah, 42.3 percent of respondents indicated that "Having enough places to ride" was most important; 8.4 percent indicated "Too many areas closed to OHV use;" and 5.6 percent indicated that "Resource management conservation" was the most important issue (Fisher and others, 2001). In a survey conducted by Nelson and others (2000), 44.6 percent of respondents selected "Do not reduce current trail/route system and OHV access" to indicate the most important thing that should not be changed, and 30.1 percent selected "Develop more trails/routes/area and connections to services" to indicate the most important thing that should be changed. When provided with several OHV-management statements with which to agree or disagree, Crimmins (1999) found that "Most trail closures have been done for good reason" received the highest level of disagreement.

Similar patterns in attitudes and beliefs were revealed through a survey of 336 ATV and motorbike users conducted by the Idaho Department of Parks and Recreation (Achana, 2005). On a scale of 0-7 (from least to most serious), respondents were asked to rank 23 issues of concern to them. Results indicated that the most serious issues of concern (in descending order of seriousness; scores greater than 4) were

- permanent closure of an area the recreator uses most,
- temporary closure of an area the recreator uses most,
- inattentive/careless recreators engaged in motorized recreation,
- litter,
- too many rules and regulations, and
- poor communication of rules and regulations.

Conversely, respondents felt that issues they were not were not concerned with (in ascending order of seriousness; scores less than 3) were

- too few rules and regulations,
- inadequate facilities at campsites,
- ATV impacts on water,
- motorcycle impacts on water,
- problems with parking availability for OHV-support vehicles,
- lack of suitable campsites,
- ATV impacts on wildlife, and
- some other (unlisted) issue of concern in OHV use areas.

Issues of concern that fell in the middle (in descending order of seriousness) were

- inattentive/careless non-motorized recreators,
- OHVs traveling too fast,

- motorcycle impacts on soil,
- motorcycle impacts on vegetation,
- ATV impacts on vegetation,
- hunters on OHVs off designated roadways and trails,
- ATV impacts on soil,
- motorcycle impacts on wildlife, and
- noise from OHVs.

When asked which of 16 possible factors contributed to creation of unauthorized trails in recreational regions of Idaho, survey respondents selected (from most to least frequently)

- belief that OHV users should be free to go anywhere,
- lack of enough designated places to ride,
- avoidance of crowded designated areas,
- lack of operator experience,
- riding motorcycles for fun,
- treeless terrain,
- riding ATVs for fun,
- using ATVs for hunting access,
- lack of enforcement regulations,
- using motorcycles for hunting access,
- using ATVs for camping,
- inadequate regulation,
- using motorcycles for camping,
- using motorcycles for fishing access,
- using ATVs for fishing access, and
- some other (unlisted) regional resource impact.

Combined, the top three possible factors contributing to creation of unauthorized trails indicate that closures of OHV areas could result in at least local increases in dispersed use. Finally, when presented with a list of four alternatives for creating uniform OHV access requirements to all recreation areas, trails, and roads on Idaho public lands, 53 percent of the respondents selected the alternative "Open to OHVs unless posted as closed by signing," and 33 percent selected the alternative "Open to OHVs unless posted as closed by signing, designation, or description." Only 6.1 and 1.0 percent felt that areas should be "Closed to OHVs unless open by signing, designation, or description" or "Closed to OHVs unless open by signing," respectively (6.7 percent did not respond to this question). These results are consistent with the top possible factors contributing to creation of unauthorized trails: the belief that OHV users should be free to go anywhere unless posted as closed by signing, designation, or description.

2.7.5. Economic Benefits and Costs of OHV Use

The economic benefits resulting from OHV sales, operation and maintenance, and associated sales and activities have been well documented (American Motorcyclist Association, 1978; Dave Miller Associates, 1981; Reed and Hass, 1989; Dean Runyon Associates, 2000; Nelson and others, 2000). OHV recreation and camping, in particular, can generate significant revenues for local economies through campground fees, grocery sales, eating and drinking in restaurants, and sales associated with operating and maintaining OHVs and support vehicles. In 1999, camping at public campgrounds on local, State, BLM, and U.S. Forest Service lands in California generated $500 million; an additional $130 million was spent solely on going to and from the campground and/or home (Dean Runyon Associates, 2000). A study conducted in 1988 by Reed and Hass (1989) indicated that, during a 12-month period in 1987-1988, Colorado OHV users spent $488.7 million on OHV purchases, operation and maintenance, support equipment (tow trailers, storage sheds, and so on), and travel expenses associated with OHV trips. Nelson and others (2000) reported that, between July 1998 and June 1999, the average Michigan OHV licensee spent $1,944 on non-trip related purchases, 80 percent of which was for equipment. When extrapolated to the estimated number of licensees in Michigan, Nelson and others (2000) found that this amounted to $134 million in spending on equipment; a similar extrapolation indicated that $40 million was spent on local trips.

The literature search conducted for this report, as well as personal communications with experts working in the field of outdoor recreation socioeconomics, revealed no published studies on the socioeconomic costs generated by OHV use. These costs could include the degradation or loss of ecosystem services, the costs of restoring OHV sites, and the loss of revenues from non- motorized recreators who seek alternate areas for recreation where motorized recreation does not occur. Examples of degraded or lost ecosystem services would be the diminished capacity for a given watershed to provide high-quality water, diminished water infiltration into aquifers, and flooding resulting from increased runoff where soils become compacted. Lost constituencies (and associated revenues) could include not only non-motorized recreators, but also hunters and anglers whose primary recreational foci (wildlife and fish) may have undergone population declines due to the effects of OHV use. At this time, however, the true benefit:cost ratio of OHV use remains unknown.

2.7.6. Annotated Bibliography for Socioeconomic Implications of OHV Use

Badaracco, R.J., 1976, ORVs—Often rough on visitors: Parks and Recreation, v. 11, no. 9, p. 32-35, 68-75.

This paper first reviews relevant literature on user conflict and discusses the one-sidedness of conflicts between OHV and non-OHV users, as well as the spatial nature of conflicts that occur when non-OHV users seek solitude and quiet and OHV users seek places for challenge and adventure. The paper then describes the ISD (impairment, suppression, displacement) syndrome: impairment is the diminished enjoyment among non-OHV users when they come into direct or indirect contact with OHV impacts; suppression is reduced participation of the non-OHV group; and displacement is the abandonment of a site impacted by OHV activity. Land planners and managers often misinterpret displacement as disinterest in the abandoned activity and, in so doing, may focus management efforts and other resources on OHV user demands.

Bury, R.L., and Fillmore, E.R., 1974, Design of motorcycle areas near campgrounds—Effects on riders and non riders: College Station, Texas, Department of Recreation and Parks, Texas A & M University, Technical Report, 72 p.

This document analyzes some of the psychological and sociological effects of constructing motorcycle riding areas adjacent to fixed-site campgrounds. It describes rider and camper profiles, rider and camper perceptions of riders, and camper and rider preferences and satisfactions with respect to the proximity and design of riding areas.

Cordell, H.K., Betz, C.J., Green, G., and Owens, M., 2005, Off-highway vehicle recreation in the United States, regions, and states—A national report from the National Survey on Recreation and the Environment (NSRE): U.S. Forest Service, Southern Research Station, Technical Report, 90 p.

This report was prepared for the U.S. Forest Service's National OHV Policy and Implementation Teams. The data from the NSRE were collected between the fall of 1999 and late 2004. The focus of this report is off-highway driving of motor vehicles. The 15 July 2004, U.S. Forest Service draft rule regarding management of motorized vehicle use has increased attention on where and how OHV recreation occurs and is offered. As public land managers are tasked with the responsibility of examining and implementing clear and consistent agency policy, understanding who the OHV recreators are has become ever more important. The growing use of motor vehicles is prompting the Forest Service to revise its management of this use so that the agency can continue to provide opportunities desired by the public, while sustaining National Forest System lands.

Crimmins, T., 1999, Colorado off-highway vehicle user survey—Summary of results: Denver, Colorado, Colorado State Parks, Technical Report.

This report summarizes a State Parks user survey designed to elucidate OHV rider-use patterns, what riders want in a recreation area, enthusiast values and beliefs, use of OHVs in hunting, how the state OHV fund should use the funds collected, and rider perceptions of how OHV funds are used, lands are allocated, and routes are managed.

Dave Miller Associates, 1981, An economic/social assessment of snowmobiling in Maine: Windham, Maine, Dave Miller Associates, Technical Report, 52 p.

This summarizes a user survey covering economics (number of trips, distance traveled, duration, fuel, lodging, equipment) and analyzing the statewide impacts and trends indicated by the responses. (No information on demographics or user perception was gathered.)

Dean Runyan Associates, 2000, Campers in California—Travel patterns and economic impacts: Portland, Oregon, Dean Runyan Associates, Technical Report, 76 p.

This document charts the distribution of camping opportunity according to type of environment and land ownership, tallies the results of a questionnaire distributed to people using public campgrounds, and develops a comprehensive profile of camping travel patterns, demographics, and expenditures. The report provides significant detail on a wide range of camping patterns, such as how many trips, how long and where, a breakdown of the activities pursued by campers once on site, and the ethnic and income classifications of campers. Although not OHV-specific, it shows where OHV recreation fits into the big picture.

Decker, D.J., Krueger, R.A., Bauer, Jr., R.A., Knuth, B.A., and Richmond, M.E., 1996, From clients to stakeholders—A philosophical shift for fish and wildlife management: Human Dimensions of Wildlife, v. 1, no. 1, p. 70-82.

This paper begins with a call for wildlife professionals to "adopt and use the term stakeholder," the development of which they review and the definition of which they indicate as being any citizen potentially affected by or having a vested interest in an issue, program, action, or decision leading to an action. The authors maintain that successful natural resource management in today's society requires recognizing the array of stakeholders that demand a voice or involvement in decision-making about natural resource management. The authors describe taking a stakeholder approach to planning and decision-making in natural resource management by including all those who might be impacted by natural resource management decisions (the authors focus on fish and wildlife management, but the principle is applied throughout natural resource management). The process entails developing communication strategies for understanding and representing stakeholder concerns, attitudes, and conflicts. The authors maintain that today's successful professional resource managers need to "...seek a widely recognized image of giving unprejudiced consideration to all significant stakeholder interests in management decisions."

Fisher, A.L., Blahna, D.J., and Bahr, R., 2001, Off-highway vehicle uses and owner preferences in Utah: Logan, Utah, Institute for Outdoor Recreation and Tourism, Department of Forest Resources, Utah State University, Report no. IORT PR2001–02, 80 p.

This study entailed an OHV user survey to examine owner characteristics, attitudes, and preferences. Respondents were selected at random from Utah OHV registrations and interviewed by telephone. This was a very extensive questionnaire, including the verbatim responses to interviewers' open-ended questions. Other questions included demographics, vehicle type used, where ridden, distance traveled, types of riding preferred, attitudes toward OHV program fund use, attitudes toward training and safety, and much more.

Jim, C., 1989, Visitor management in recreation areas: Environmental Conservation, v. 16, no. 1, p. 19–32.

This paper discusses various visitor-management measures for diminishing or precluding the effects of visitor impacts on natural resources in recreation areas by employing existing recreation-management research on visitor decisions—such as trip duration, difficulty, and desired environment—to suggest ways of dispersing use into patterns that do not result in damage to natural resources. It also examines various management scenarios: signs and maps to direct users into a managed pattern, restricting admission, lotteries, and various rationing/pricing concepts.

Kockelman, W.J., 1983, Management Concepts, *in* Webb, R.H., and Wilshire, H.G., eds., Environmental effects of off-road vehicles—Impacts and management in arid regions: New York, Springer-Verlag, p. 399–446.

Noise and motorized intrusion were the major impacts of ORVs on non-OHV users. Permitting OHV activity on public land is described as "inefficient" in the goal to provide for multiple uses because the noise, dust, and speed of just one OHV can exclude all other recreators from an area. The author categorizes OHV users as work-related users, recreational users, or "bad apples." Work-related users are natural resource managers and utility workers, among others. Recreators are further categorized as casual (value aesthetics more than the challenges of riding) or endurance riders. "Bad apples" are characterized by a complete lack of concern about their impacts and are likely to be noncompliant with regulations.

Nelson, C.M., and Lynch, J.A., 2001, A usable pilot off-road vehicle project evaluation: East Lansing, Michigan, Department of Park, Recreation and Tourism Resources, Michigan State University, Technical Report, 50 p.

This report details the results of an interagency effort to increase compliance with OHV rules in a Michigan State forest. An OHV-rider survey asked for respondents' perceptions of signs, maps, and trail systems in the pilot area, as well as rider perceptions of any law enforcement contact riders may have had during the study period. The survey also queried each respondent's understanding of pilot area regulations and offered the opportunity to give open- ended comments. There is also a detailed discussion of the participating law enforcement agencies' response to the pilot project, including officer concerns, jurisdiction conflicts, workload distribution vs. agency priorities, and an analysis of sign survival in the pilot project areas. Finally, interviews with park manager/grant recipients and discussion of the results in terms of park administration, funding, staffing, and resource protection are provided.

Nelson, C.M., Lynch, J.A., and Stynes, D.J., 2000, Michigan licensed off-road vehicle use and users 1998–99: East Lansing, Michigan, Department of Park, Recreation and Tourism Resources, Michigan State University, Technical Report, 49 p.

This details a survey of randomly selected OHV owners in 1999. In addition to questions about demographics, expenditures, type of OHVs owned, and preferred activities, respondents were queried about their perceptions of specific aspects of the State OHV program. One section is dedicated to comparing this survey with a similar survey from 1988.

Propst, D.B., Shomaker, J.H., and Mitchekkm, J.E., 1977, Attitudes of Idaho off-road vehicle users and mangers: Moscow, Idaho, College of Forestry, Wildlife and Range Sciences, University of Idaho, Technical Report, 30 p.

This report provides background information on, and an introduction to OHV use in, the era when it was new and poorly understood, and includes one of the earliest OHV/OSV (over-snow vehicle) user surveys. It compares user and land manager responses in the same survey; both groups were queried about their perceptions of environmental impacts, causes of conflicts, uses of public money for facilities, regulation enforcement, impacts on wildlife, and reasons for pursuing OHV/OSV activities.

Chapter 3

POTENTIAL INDICATORS FOR EVALUATING AND MONITORING OHV EFFECTS

3.1. SUMMARY

There are numerous parameters that have the potential for serving as indicators of OHV effects in monitoring or research programs. Every attempt was made to provide an inclusive list of potential indicators of OHV effects described in the OHV effects literature (listed below). Of those listed, some correspond with BLM's 17 indicators of rangeland health; others are quite different but could provide supplemental data for evaluating or monitoring OHV effects (for example, erosion and/or sedimentation rates would complement assessments of rill formation and other surface changes) or fill indicator voids (such as those pertaining to wildlife ecology).

(1) *Soil health* and *watershed condition*
  - Soil strength
  - Soil bulk density
  - Water infiltration rate
  - Permeability
  - Erosion and sedimentation rate
  - Sedimentation or turbidity in wetlands
  - Surface changes (for example, formation of rills, gullies, and terracettes)
  - Presence/condition of soil crusts (in some cases: depending on crust type)

(2) *Vegetation health*
  - Plant community composition (including species diversity, ratio of native to non-native or invasive species, structural diversity)
  - Abundance of individuals and/or stem density
  - Percent vegetation cover
  - Plant size
  - Growth rate
  - Biomass

(3) *Habitat condition* and *health of wildlife populations* (direct and indirect)

- Habitat patch size and connectivity
- Wildlife community composition (including species diversity, ratio of native to non-native or invasive species)
- Abundance, density, and distribution
- Population sizes and trends
- Survivorship, productivity, body mass, and roadkill rates
- Age-class and gender structure
- Frequency of OHVs passing through a given area
- Road or trail type and width
- Level (decibels), duration, and timing of traffic noise

(4) *Water quality*

- Sedimentation rate
- Levels of turbidity and suspended solids
- Contaminants levels, including levels of petroleum-derived compounds from spills (aromatic hydrocarbons in particular)

(5) *Air quality*

- Dust levels
- Levels of by-products of OHV emissions (including polycyclic aromatic hydrocarbons, carbon monoxide, nitrogen oxides, ozone, and sulfur dioxide)

(6) *Socioeconomics* (direct and indirect)

- Recreator satisfaction with their recreation (or other) experiences
- Compliance with OHV (or other) regulations
- Knowledge regarding effects of user activities on various aspects of land health
- Mapping the distribution and intensity of OHV versus non-motorized recreation and other land uses,
- Patterns of regulation compliance (as evidenced by creation of unauthorized trails, damage to vegetation, and so on)
- Trends in local economic indicators associated with OHV and non- motorized recreation and other land uses (for example, sales in camping equipment, gasoline, restaurants, lodging facilities)

Specific research questions and management goals—as well as sensitivity to OHV effects and the availability of funding and personnel—will determine the potential efficacy of using any one indicator to evaluate or monitor OHV effects on BLM lands. Qualitative indicators may be most useful for rapid assessments, whereas quantitative indicators may be needed for long-term monitoring. Ultimately, however, implementing an OHV effects monitoring program will require consultation with topical experts and additional research to identify or develop appropriate and efficient indicators and field methods for evaluating and monitoring OHV effects (personal communication from D.A. Pyke to Z.H. Bowen, U.S. Geological Survey, Fort Collins, Colorado, August 2007). Work on developing such indicators is currently underway by rangeland ecologist, D.A. Pyke, U.S. Geological Survey in Corvallis, Oregon.

3.2. BLM's Indicators of Land Health Compared to Indicators of OHV Effects Described in the Literature

In terms of the specific land health attributes assessed, there is some limited correspondence between several indicators of OHV effects described in the literature and some of BLM's 17 qualitative assessment indicators (see Pellant and others, 2005). The area of greatest overlap is that of soil health and watershed condition; there is somewhat less overlap in the area of vegetation health (table 3.1). Attributes addressed in the literature but not by BLM's 17 indicators include wildlife population and habitat health, water and air quality, and socioeconomics. Even indicators that measure the same or similar attributes, however, may differ notably with respect to the scale and scope to which they are or can be applied, or the precision and accuracy they can provide.

The differences between BLM's indicators and those described in the literature do not imply that BLM's indicators are inappropriate for assessing some attributes under some conditions. Rather, they underscore the need for a variety of indicators to meet equally variable needs. For example, qualitative indicators (such as those employed by the BLM) often entail making visual estimates, which may be suitable for rapid assessments by time-limited personnel operating with small budgets; qualitative measurements, however, are subject to observer bias. Research and monitoring studies, on the other hand, generally require quantitative indicators (or strict decision rules to guide data collection for qualitative parameters) that minimize observer bias and maximize statistical precision and accuracy to ensure defensible results and the detection of trends; quantitative measurements, however, can drive up the cost and time requirements of research and monitoring efforts. Therefore, the choice of indicators employed will depend on the specific goals, budgets, sites, and other factors. In some cases, BLM's indicators may be suitable; other cases may require more quantitative indicators. Co-opting indicators from other disciplines also may be extremely useful for revealing OHV effects on land health.

Table 3.1. Indicators emphasized in the literature reviewed for effects of off-highway vehicles (OHV) on land health compared to indicators of land health employed by the U.S. Bureau of Land Management (BLM) (Pellant and others, 2005)

Land health category	Indicators of OHV effects described in reviewed literature	BLM indicators[a]
Soils and watersheds	Soil strength	9,11*
	Soil bulk density	11
	Soil permeability	8,9*,11*
	Water infiltration rate	1,2,3,4,5,8*,9*, 10*,11*
	Erosion rate	1,2,3,4,5,6,7, 8*,10*
	Sedimentation rate	
	Presence/condition of biotic and abiotic soil crusts	2,4,6,7,8*

Table 3.1. Continued

Land health category	Indicators of OHV effects described in reviewed literature	BLM indicators[a]
Vegetation	Plant species diversity	12*,13,16*
	Ratio of native plants to non-native and/or invasive plants	12,16
	Percent plant cover	4,13,14
	Plant size	11,13,14, 15*,17
	Plant growth rate	11*,13,14,15*, 17
Wildlife and habitats	Habitat patch size and connectivity (can be expressed as ratio of road edge:habitat area or as native:non-native habitat)	4,12*,16*
	Shape/scope of animal movements relative to roads	
	Wildlife diversity and/or species abundance	
	Ratio of native, endemic wildlife to non-native and/or invasive species	
	Population size and trend	
	Gender/age ratio trend	
	Productivity trend	
	Average body mass for a given age/gender	
	Average survivorship	
	Vehicle-caused mortality rate	
Water quality	Sedimentation rate or depth	
	Amount of suspended solids	
	Turbidity level	
	Level of atmospheric deposition associated with OHV emissions	
	Level of petroleum or its by-products (benzene, ethylbenzene, toluene, xylenes, 1,3-butadiene, lead)	
Air quality	Levels of OHV emission by-products (nitrogen oxides, carbon monoxide, sulfur dioxide, ozone, aldehyde, PAHs[b])	
	Level of suspended particulates	
	Plant-absorbed level of emissions by-products	

[a] 1 = Number and extent of rills.
2 = Presence of water flow patterns.
3 = Number and height of erosional pedestals or terracettes.
4 = Bare ground (excluding rock, litter, lichen, moss, plant canopy).
5 = Number of gullies and erosion associated with gullies.
6 = Extent of wind scoured blowouts and/or depositional areas.
7 = Amount of litter movement (description of size and distance expected to travel).
8 = Average soil surface (top few mm) resistance to erosion.
9 = Soil surface structure and content of soil organic matter (to include type of structure and A-horizon color and thickness).

10 = Effect of plant community composition (relative proportion of different functional groups) and spatial distribution on infiltration and runoff.
11 = Presence and thickness of compaction layer (usually none).
12 = Functional/structural groups (in descending order of dominance by above-ground production or live foliar cover).
13 = Amount of plant mortality and decadence (include which functional groups are expected to show mortality or decadence).
14 = Average percent and depth of litter cover.
15 = Expected annual production (this it total above-ground production, not just forage production).
16 = Potential invasive (including noxious) species (native and non-native).
17 = Perennial plant reproductive capability.
[b] Polycyclic aromatic hydrocarbons.

Ultimately, *any indicator used to evaluate or monitor OHV effects will need a standard value (or range of values) that represents the baseline condition or a threshold value above (or below) which management is triggered.* Ecosystem properties, however, can vary widely across (and within) *spatial scales*. Therefore, selecting an appropriate standard or threshold for any one indicator necessitates an evaluation of the spatial scale(s) at which the associated effect occurs or is likely to affect ecosystem function. For example, OHV-caused sedimentation in a first-order stream might have significant by very localized effects on that stream, but no significant effects downstream; in contrast, OHV-caused sedimentation in a first- order stream might not have significant effects on that stream, but sedimentation in multiple first-order streams may result in significant cumulative downstream effects in second- and third- order streams. Similarly, the *temporal scale* at which ecosystems respond to land uses can vary from short-term (hours) to long-term (decades or longer); thus, it is important to consider temporal scale(s) as indicator standards or thresholds are established. For example, wildlife behaviors may exhibit immediate responses to OHV traffic by moving away from OHV routes; population trends, however, may take several generations to show effects of OHV disturbance, in which case the time required to detect trends will depend on how long it takes for generations turn over. Where OHV effects are not an immediate concern, one could incorporate long-term, annual, qualitative assessments of OHV-use areas, but where effects are of immediate concern, short-term, quantitative assessments may be implemented.

Standards for a given land health indicator can be developed by ascertaining baseline values at control sites under a given set of conditions. Once the baseline values are established, it becomes important to standardize the conditions under which subsequent measurements of that indicator are made. For example, the freezing and thawing of soils tend to decompact soils, and precipitation tends to alter other soil properties; thus, it would be important to monitor soils at similar times of year and under similar soil-moisture conditions.

3.3. Some Potential Indicators for Evaluating and Monitoring OHV Effects

3.3.1. Potential Indicators of OHV Effects on Soils and Watersheds

A major effect of OHV traffic on soil health is compaction—the reduction of a soil's porosity. Potentially useful indicators for monitoring soil compaction include soil strength, soil bulk density, and infiltration rate (or permeability). *Soil strength* is typically measured in terms of the soil surface's resistance to a vertical force exerted by a penetrometer and is expressed as kg/cm^2 (see *http://cropsoil.psu.edu/extension/facts/agfacts63.cfm* for explanations and diagrams of this method). *Soil bulk density* is measured as the ratio of dry solid mass (after soil is over- dried) to bulk volume of soil and is expressed as g/cm^3 (or Mg/m^3). *Infiltration* or *permeability* is the rate at which the water infiltrates the soil and is expressed as cm/hr. Infiltration tests can be conducted in the field with relatively simple equipment (for a demonstration, see *http://www.grow.arizona.edu/Grow--GrowResources .php?Resourceld=181);* however, many replicates are needed to obtain adequate comparisons, and it takes time for infiltration to occur. Additional resources on the utility of devices and techniques for monitoring soil health include Leung and Meyer (2003) for soil compaction; O'Sullivan and Ball (1982), Hooks and Jansen (1986), Komatsu and others (1988), and Keener and others (1991) for soil strength; and Flint and Childs (1984), Isensee and Luth (1992), Miller and others (2001), and Lowery and Morrison (2002) for soil bulk density. McBrayer and others (1997) and Amezketa Lizarraga and others (2002) describe techniques used to measure infiltration rates. Overall, soil strength and bulk density are the most commonly used indicators of soil compaction in visitor impact studies (Liddle, 1997).

Soil strength, which increases with increasing soil compaction, depends on a number of inherent variables, such as particle size, type of clay mineral, the size/distribution of pores, and aggregate stability. For example, Adams and others (1982) found that soil strength in undisturbed areas of the Mojave Desert ranged between 5.1 kg/cm^2 at 6 percent water content and 21.1 kg/cm^2 at 1.8 percent water content, indicating that soil strength decreases with increasing moisture content. In general, a soil strength of less than 20 kg/cm^2 (284.4 lb/in^2) was indicative of undisturbed terrain, whereas trails intensely used by motorcycles had soil strengths (during wet conditions) that typically ranged from 20 to 60 kg/cm^2 (284.4 to 853.2 lb/in^2) (Adams and others, 1982). When soil strengths exceeded 20 kg/cm^2 (284.4 lb/in^2; measured at about field capacity) due to compaction, Grimes and others (1975, 1978) found that root extension of certain plants, including alfalfa (*Medicago sativa*), corn (*Zea mays*), and cotton (*Gossypium hirsutum*), was limited.

Soil bulk density also increases with increasing soil compaction. For example, in the Mojave Desert, Webb (2002) and Caldwell and others (2006) found that bulk densities of undisturbed surface soils ranged from 1.40 to 1.68 g/cm^3; however, bulk density increased significantly to 1.80 g/cm^3 at high-disturbance sites (Caldwell and others, 2006). Measuring soil bulk density, however, is time-consuming and difficult in gravelly soils (Webb, 2002), which are typical of many desert sites. Thus, its practicality may vary not only from region to region, but from soil to soil within a region. Infiltration, on the other hand, decreases with increasing soil compaction. In undisturbed desert habitats, Eckert and others (1979) found soil infiltration rates to be 3.2 cm/hr. In the same study, infiltration rates were 15 percent lower where motorcycle (2.7 cm/hr) and 33 percent where truck (2.1 cm/hr) traffic had occurred.

For each of the indicators identified above, it will be crucial to consider soil type and water content when setting standards, as these factors clearly influence overall inherent values of each parameter. It also is clear that variability in soil type may preclude using a single indicator—much less one standard—for monitoring soil health across all sites. Overall, monitoring soil health would entail measuring soil properties in tracked versus untracked areas of similar soil type and under similar conditions of soil water content. For areas previously unaffected by OHV activities but proposed to become OHV areas, managers may wish to collect predisturbance data on soil properties to serve as a baseline from which trends in soil properties may be assessed over time. An acceptable percent change in soil properties could be selected, and if soil properties were to exceed these threshold values then management actions could be implemented to bring the soil properties back to acceptable values.

A possible method of monitoring surface changes in watersheds might be to establish permanent monitoring sites for repeat photography studies. By standing in the same spot, orienting the camera in the same direction, and using the same focal length for each site visit, sequential photographs would provide a qualitative, relatively easy means of monitoring surface changes due to erosion and OHV tire cuts. Helpful websites that discuss repeat photography (both terrestrial and aerial) include *http://biology.usgs.gov/luhna/chap9.html, http://wwwpaztcn.wr.usgs.gov/wyoming/rpt_ground.html,* and *http://www.cpluhna. nau.edu/ Tools/repeatphotog.htm.* Scientists with the National Science and Technology Center are using close-range (terrestrial or ground-based) photogrammetry techniques and associated tools for producing and interpreting close-range images (less than 300 m, as opposed to aerial photogrammetry distances of more than 300 m) that may be used to document soil crusts, soil erosion, vegetation, and other resources (see *http://www.blm.gov/nstc/prodserv/ ST134/pdf/ Handout3CloseRange.pdf).*

3.3.2. Potential Indicators of OHV Effects on Vegetation

Some plant responses to OHV activities are preceded by, and result from, OHV effects on soil properties. In other words, soil characteristics, particularly soil compaction, play important roles in the distribution, abundance, growth rate, reproduction, and size of plants. Furthermore, some plant responses are likely to lag behind changes in soil properties, and, by the time effects are detected in plants, site recovery could be more difficult and/or lengthy. As such, it would be important to implement management strategies for maintaining or improving soil condition before plants are affected and no longer provide enough cover to hold soils in place during restoration efforts. Several vegetation parameters, however, have potential value as direct indicators of OHV effects on plant communities.

As mentioned in section 3.3.1, soil compaction results in reduced water infiltration. As a result, overall plant productivity, as measured by *percent plant cover*, *abundance* or *stem density*, *growth rate*, *biomass*, *plant height* and *width*, *ratio of large:small species*, and/or *reproductive output* may be diminished (Johnson and others, 1975; Vasek and others, 1975; Adams and others, 1982; Webb, 1983; Prose and others, 1987; Holzapfel and Schmidt, 1990; Lightfoot and Whitford, 1991; Brooks, 1995; Bolling and Walker, 2000). It is important to consider, however, the differential growth habits and responses of plant species to conditions generated by OHV activities; they may be favored or inhibited by OHV effects (Holzapfel and Schmidt, 1990; Angold, 1997). For example, the productivity of some species may increase due to abnormal conditions of moisture availability from runoff near compacted areas and/or where water roadbed materials allow increased infiltration rates (Johnson and others, 1975; Vasek and others, 1975; Holzapfel and Schmidt, 1990; Lightfoot and Whitford, 1991).

To evaluate and monitor plant productivity, researchers often use transect-intercept methods for measuring larger- or site-scale indicators, such as percent cover, and quadrat-based sub-sampling at random locations along transects for measuring smaller-scale (individual plants) indicators, such as plant size or growth rate. Repeat photography is also potentially useful for monitoring vegetation cover when personnel budgets are limiting, although ground-based repeat photography methods may be more realistic than satellite imaging in terms of staff expertise and funding required (see methods and URLs provided in Section 3.3.1). Satellite imagery, however, can be very effective for ascertaining landscape-scale changes in vegetation cover during longterm studies at selected study sites (Johansen and others, 2007).

Plant community composition or diversity—a parameter that factors relative proportions of each species into species richness—is a commonly used indicator health of a vegetation community (Davidson and Fox, 1974; Adams and others, 1982; Prose and others, 1987; Wilcox, 1989; Tyser and Worley, 1992; Lovich and Bainbridge, 1999; Parendes and Jones, 2000). For example, a plant community comprising 5 individuals each of 10 species and 150 individuals of another (200 total individuals) would be considered depauperate compared to a community comprising 20 individuals each of 10 species (200 total). Native species diversity is further compromised when non-native and/or invasive plants dominate the plant community (Holzapfel and Schmidt, 1990). OHVs caked with mud acquired elsewhere potentially introduce or disperse seeds of non-native and invasive species; thus, OHV-route margins often become populated with exotics and invasives that eventually may spread and outcompete native species at the landscape level. Therefore, an important consideration when evaluating plant species diversity is the presence of non-native and/or invasive species. In other words, although species diversity can be useful for evaluating and monitoring the impacts of OHV activities on vegetation, the ratio of native to non-native and invasive plant species must be taken into account. Monitoring transects oriented perpendicular to OHV travel routes would help identify range expansions beyond the linear routes.

Similar to the precautions issued above for selecting standards of soil health, vegetation characteristics also vary widely according to soil type, slope, aspect, microclimate, and other factors. Ultimately, having baseline data before a site is disturbed, and/or having nearby reference sites not subjected to OHV activity would help differentiate between site-based variations and OHV impacts on vegetation.

3.3.3. Potential Indicators of OHV Effects on Wildlife and Habitats: Native, Threatened, and Endangered Species

The physical imprint of a road or OHV route creates barrier effects that may effectively alter *habitat patch size and connectivity*, which potentially alters or inhibits *animal movements* (Oxley and others, 1974; Mader, 1984; Swihart and Slade, 1984; Samways, 1989; Andrews, 1990; Baur and Baur, 1990; Forman and Alexander, 1998; Jackson and Griffen, 1998). Roads and trails also create edges that can alter wildlife habitat use and movements (Nicholson, 1987; Yahner, 1988; Reed and others, 1996; von Seckendorff Hoff and Marlow, 1997; Gibbs, 1998; Vos and Chardon, 1998), and in many instances these *edge effects* extend well beyond the road's actual footprint into habitat interiors. Unfortunately, both direct and indirect indicators of animal movements and habitat or home-range use can be difficult to measure, simply because populations are mobile and easily affected by many factors besides OHV activity if care is not taken to avoid, or control for, these additional effects. Furthermore, adequate sample sizes and accurate measurements of animal movements generally require capture-mark-release or capturemark-resight studies, which can be costly in terms of funding for equipment and personnel time. Therefore, evaluating and monitoring movement responses of suitable indicator species or functional groups to OHV activities and routes might be more efficient than trying to monitor many species. For example, species least likely to cross OHV routes or networks of routes (beetles or small mammals, for example) would be more suitable for studying barrier effects of OHV routes than species inhibited from crossing only multi-lane, paved highways (large ungulates). Ultimately, animal movement studies might be most appropriate in comprehensive, long-term research projects at selected study sites representative of locations and habitats being impacted most by OHV activities. An appropriate measure of edge effects on wildlife may entail studies of spatial distribution relative to distance from OHV routes, as well as density and other population dynamics among target indicator species that require habitat interiors.

Traffic intensity and noise level of OHVs may be useful indirect measures of OHV effects on wildlife behavior and survivorship. If measured, temporal variation in animal activity or presence would need to be considered, as diurnal animals may be more affected by OHV traffic and noise than nocturnal animals if most OHV activity occurs during daylight hours (Ouren and Watts, 2005). Vehicle speed also may be an important parameter to measure, as it can affect mortality rates of animals as they attempt to navigate landscapes where OHV travel is significant. There are technologies available for field monitoring of noise levels and traffic volume and speed (see examples on the Internet at *http://www.jhuapl.edu/ott/technologies/technology/articles/P01254.asp* and *http://www.noisemeters.com/accessories/outdoorkits.asp),* although again it would be more cost-effective to use them in selected study sites for targeted research questions as opposed to broad- scale monitoring programs.

Ideally, any program for monitoring OHV effects on wildlife would include assessments of whether/how wildlife responds to OHV-related factors. If population dynamics were understood *before* the onset of OHV activities, then monitoring any changes that occur afterward could be straightforward if other conditions are held relatively constant. Climate, however, typically makes long-term monitoring of animal population dynamics very complex, and wide- ranging animals may be more difficult to monitor than those with smaller or more restricted movements. Similar to the dynamic ways in which wildlife populations use a given area, OHV use in a given area is also dynamic; if the area affected and the intensity of disturbance enlarges considerably, wildlife may respond as well. These and other factors typically drive up the funding and staffing needs associated with collecting population dynamics data. Thus, to determine thresholds of OHV activity potentially tolerated by wildlife, it may be more realistic to measure long-term changes relative to gradients and changes in OHV activity. Again, careful selection of suitable indicator species, or functional groups, may help increase the effectiveness of monitoring wildlife populations (Lindenmayer, 2000; Noon, 2003, p. 5 1-55).

To date, there have been few simultaneous studies of OHV use and wildlife responses to OHV activities, but such studies could provide more precise assessments of the relationship between patterns of OHV activity and wildlife responses, particularly behavioral responses to varying traffic patterns, intensity, and total area affected. GPS and satellite technologies could be employed to build Geographic Information System (GIS) data layers of OHV-associated variables, both static (for example, road width) and dynamic (vehicle speed and traffic volume). When overlaid with GPS-based telemetry data layers that map animal movements, one could relate changes in static and dynamic OHV variables to wildlife responses, making these particularly powerful tools for long-term studies aimed at evaluating OHV impacts on wildlife.

3.3.4. Potential Indicators of OHV Effects on Water Quality

The literature on OilV impacts offered little information specific to evaluating or monitoring OilV effects on water or air quality. Based on studies of water quality in other disciplines, however, potentially useful indicators highly relevant to OilV effects would be *sedimentation* and *turbidity*. Sedimentation can be measured in terms of deposition rate or total amount of solids deposited where surface and directed flows enter aquatic systems downslope of OilV-use areas. A useful reference on different methods for measuring sedimentation may be found in Lisle and Eads (1991). Turbidity, which indicates the level of *suspended solids* in water, can be measured easily in the field with a Secchi disk (see *http://www.noble.org/Ag/Wildlife/SecchiDisk/Index.htm)*. With respect to monitoring levels of contaminants in water from OilV emissions and fuel or other chemical spills, water samples can be collected and analyzed in laboratory settings; however, this can be costly, and would probably be more suitable for selected, long-term research sites than in broad-scale monitoring programs. Potentially important contaminants to test for in OilV-impacted watersheds could include *benzene*; *ethylbenzene*; *m-, p-, and o-xylene*; *toluene*; *1,3-butadiene*; *and lead* (Forman and others, 2003: p. 205-213; Arnold and Koel, 2006). In addition, nitrogen deposition from nitrogen oxides can affect water quality if nitrogen loading alters the chemical balance of nutrients in aquatic organisms.

Although it has already been stated that, in general, it is important to compare OilV-impacted sites with similar reference (unaffected) sites when evaluating OilV effects, an additional consideration with respect to water quality is to make comparisons within the same drainage. This is because water quality variables can change depending on the geology, adjacent habitat, and hydrology of the area. Thus, an appropriate technique for assessing OilV effects on aquatic systems might be to compare water quality at replicate sites both upstream and downstream of where OilV activities are occurring.

3.3.5. Potential Indicators of OHV Effects on Air Quality

There are several measures of air quality that can be used to assess effects of OilVs, including levels of fugitive dust (*suspended particulates*) and/or *OHV emissions* (including *carbon monoxide*, *ozone*, *sulfur dioxide*, *aldehyde*, and *polycyclic aromatic hydrocarbons*). Due to the effects of humidity, precipitation, fallout rates of different particle sizes, and wind speed and direction, however, measuring dust levels specific to any one site or set of OilV activities can be difficult. A useful technique for assessing the amount of dust associated with OilV use is to collect PM10 (particulate matter less than 10 microns in diameter, which can pass through the nose and throat and get deep into the lungs) data, as dust is a common component of PM10. Technological advancements continue to provide additional devices useful for monitoring dust and other suspended particulates (Sanders and Addo, 2000)—including satellite imagery (Nakata and others, 1976; Gill, 1996; Stefanov and others, 2001)—although cost becomes a greater factor with increasing sophistication of instruments used. Fox (1986) provides a list of appropriate procedures for measuring and evaluating various air-quality parameters, and the Environmental Protection Agency's National Ambient Air Quality Standards include primary standards (those designed to protect overall public health and the health of "sensitive populations") for carbon monoxide, lead, nitrogen dioxide, particulate matter, ozone, and sulfur oxide levels (see *http://www.epa.gov/air/criteria.html).*

3.3.6. Potential Indicators for OHV Effects on Socioeconomics

Resource planning has been known to take recreation and economic values of OHV use into greater consideration than biological considerations (Adams and Dove, 1989). *Human behaviors, attitudes, and economics*, however, are the ultimate drivers of OHV impacts on natural resources (Decker and others, 1996; Vaske and others, 2001); thus, understanding the socioeconomics of OHV use is crucial to the success of any program designed to address OHV effects, whether they impact natural resources or land users. This includes understanding the economic effects—*both benefits and costs*—of OHV management and use on OHV users, other land users, local businesses, land-management agencies, and ecosystem services.

Bight and others (2003) provide a framework and guidelines for conducting social assessments and identifying/organizing social science data, including measurable indicators (see Chapter 3, p. 21) for use in natural resource planning. Indicators suitable for monitoring the socioeconomic implications of OHV use (human behaviors, attitudes, and economics) may be identified through stakeholder interviews, focus groups, and surveys (mail or telephone); economic assessments; and developing maps depicting areas used for different forms and intensities of recreation (Massachusetts Department of Environmental Management, 1995; Decker and others, 1996; Stokowski and LaPointe, 2000; Nelson and Lynch, 2001; Dillman, 2007). For example, interviews can be designed to elucidate not only OHV impacts on all types of user experiences, but also ways in which those impacts might be mitigated. Trail openings or closures may affect levels of OHV user demand and satisfaction, which could be identified through well-designed survey questions that target OHV users (see Dillman 2007).

Likewise, understanding the effects of OHV site development and regulation compliance first requires monitoring where OHV users like to go and what their preferences are when riding. For example, Nelson and Lynch (2001) conducted a study to determine the effectiveness of OHV areas, the approaches for which included a survey of licensed OHV riders, interviews with key stakeholders about project management, and an assessment of the signs established to identify designated OHV trails. Monitored over time, these indicators could be used as guidelines for adaptive management. Finally, identifying the economic impacts associated with OHV activities and regulations potentially affecting nearby communities may entail economic analyses of businesses and services that cater to outdoor recreators (including OHV users and non-motorized recreators; English and others, 2001). Economic assessments also may be used to determine the financial effects associated with losing ecosystem services, regulation enforcement, resource restoration, and other potential costs of OHV use.

A potentially powerful tool for identifying socioeconomic effects of OHV use would be GIS applications (Massachusetts Department of Environmental Management, 1995; Stokowski and LaPointe, 2000; Kopperoinen and others, 2004). For example, areas considered by recreators as being most important for excluding motorized forms of recreation could be identified by users on a map and then used to develop time-series GIS data layers that illustrate long-term changes in preferences pertaining to land-management actions or other factors. Overlaid with maps that identify areas ecologically most suitable for OHV use, managers also could determine which areas are most likely to be resilient to OHV use and provide OHV user satisfaction.

Chapter 4

MITIGATION AND SITE-RESTORATION TECHNIQUES

4.1. SUMMARY

Mitigation of OHV effects and restoration of OHV-impacted sites requires a range of approaches and techniques. Social science in particular has strong applicability for ameliorating the effects of OHV activities, not only in terms of their impacts on non-motorized recreators and other OHV users, but also in terms of their effects on natural resources, as ultimately human behavior is what drives OHV use and related behaviors. Important tools for managing OHV use, therefore, include not only interviews, surveys, and focus groups, but also strong educational campaigns. Once impacted by OHV use, however, ecosystems may need to be closed and rested, if not restored. Sites with severely compacted soils and/or bedrock exposures due to erosion may need restoration through importation of native soils, scarification, decompaction, stabilization, inoculation with microbes and mycorrhizae, and/or mulching before reseeding and/or planting can be done.

4.2. MITIGATION AND SITE-RESTORATION TECHNIQUES

The OHV literature has addressed many effects of both motorized and non-motorized recreation on components of ecosystem health (Cole, 2004; Stokowski and LaPointe, 2000; Cline and others, 2007), including soils, vegetation and habitat, wildlife behavior and population dynamics, and the quality of water and air. Resource degradation and wildlife disturbance are not uniform, however. Factors influencing the extent and degree of impact include user types and behaviors, the environment's resistance and resilience, and the timing, intensity, and distribution of use (Cole, 2004). Therefore, management and mitigation planning and implementation must take these factors into consideration.

4.2.1. Understanding Land User Preferences and Conflicts

Although addressing ecosystem degradation and user conflicts stemming from OHV use generally requires policy and management considerations (Vancini, 1989), there is the likelihood that one group or another will be dissatisfied with the outcome of any one management decision. Therefore, it is important to promote acceptance of, if not support for, management decisions prior to implementation, because those not accepted are likely to fail in the long run, regardless of how sound the reasoning is behind them (Shindler and others, 2004). One way to potentially improve policy acceptability is to identify, assess, and include all users who may be affected by management decisions and include them in the decision-making process; useful tools for accomplishing this goal include stakeholder and demographic analyses, and arranging stakeholder focus groups (Decker and others, 1996; National Oceanic and Atmospheric Administration, 2005). Basic discussions, such as defining different interpretations of OHV use or seeking consensus on the meaning of OHV use to disparate users, can potentially preclude feelings of marginalization by any given group that could lead to conflict between users and/or between users and managers (Stokowski and LaPointe, 2000).

A baseline understanding of OHV users (including recreators, livestock operators, and energy-development operators) can further help to alleviate conflicts among different users. If users can be classified, even broadly, managers can maximize their efforts by using that information to form relevant management plans and/or communication and education campaigns. For example, understanding the problems that users have with each other, as well as their motivations for recreating, can yield more effective management that placates most, if not all stakeholder groups, and does not lead to marginalization or displacement of any one group. Often, outdoor recreators of all types have fairly similar goals and reasons for participating in outdoor recreation (Schuett and Ostergren, 2003), including a need to "get away" from the pressures and commotion of everyday life, rejuvenation, and enjoyment of the natural environment. Even when the mechanisms by which those needs are met are not homogeneous among user groups, knowledge of stakeholder preferences can help to improve the management of OHV recreation on public lands.

Managing the social effects of OHV activities and promoting compliance with regulations also require determining where OHV users like to go and what their preferences are when riding (Nelson and Lynch, 2001). Likewise, management must consider the levels of environmental dominance and technology associated with different forms of land use. Once again, GIS data layers could be developed to identify spatial management needs and predict where conflicts may occur. For instance, nature study, an activity characterized by "low dependence on technology and low dominance over the environment," and OHV touring, which is characterized by "high dependence on technology and high levels of environmental dominance," need to be segregated spatially to help prevent user conflicts. If it is found on a map grid that two such activities are taking place in adjacent quadrats, there is a high likelihood that recreators who participate in these activities will come into conflict (Bury and others, 1983). Maps of sites suitable for a given activity could be overlaid with user preferences for certain locations or landscape features to further fine-tune planning and mitigation of user conflicts.

Finally, if recreators are made aware of their impacts and understand the implications of those impacts, they may be more willing to take steps that lessen those effects, thereby diminishing the necessary level of managerial monitoring (Vancini, 1989; Anderson and others, 1998). Communication and education campaigns can be difficult, however, due in part to (1) modern-day information overload, increasing the likelihood that recreators will disregard information pertaining to natural resources and recreation, and (2) the fact that managers and designers of education campaigns rarely have a formal knowledge and understanding of persuasion techniques (Absher and Bright, 2004). Assuming that these factors can be surmounted by well-designed educational resources and manager training, education is a crucial first step in alleviating negative perceptions of trail closures and other regulations or management actions. For example, management agencies need to explain the reasons and rationale for closures so that they do not appear arbitrary (Crimmins, 1999). Such educational campaigns may be more successful if they target the many OHV users participating in some other form of outdoor recreation that depends on a healthy ecosystem. Hunting and fishing are two such forms of recreation, and if an OHV area is closed seasonally to protect elk during calving season or fish during spawning season, communicating this to OHV users is likely to elicit more compliance with the closures. When OHV users understand that their hunting and fishing activities may be at stake, they are more likely to respect closures. In closed areas already subjected to high levels of illegal riding, education would be crucial for communicating to users the reasons for the closures and related management actions.

4.2.2. Mitigating OHV Use Effects

Lands managed by BLM are placed into one of three broad types pertaining to OHV-use designation: "open," "limited," or "closed." Within the "limited" category there can be several types of limitations: OHVs can be "limited to existing routes," "limited to designated routes," or "limited seasonally" (see *http://www.blm.gov/nhp/news/releases/pages/2000/pr000110_ohv_qa.html*). Indeed, trail/area closures are among the management options available for allowing soils and vegetation to recover from OHV effects or to help preclude localized impacts on air and water quality. However, in areas where OHV effects would be notable with the first few uses and/or generate significant, long-lasting impacts, it would be prudent to consider whether such places are appropriate for OHV use in the first place (Cole and Landres, 1995; Cole, 2004). A well-placed, concentrated system of trails could alleviate the difficulties of enforcement that extend across a large territory (Major, 1987). Spatial models, including GIS data layers, developed for identifying areas most suitable for resisting or recovering rapidly from OHV impacts, or those most suited for concentrated and/or self-monitored OHV use, would be effective tools for establishing OHV sites in appropriate areas and avoiding the need for frequent, long-term closures, expensive restoration actions, and/or the high cost or enforcement monitoring. For example, erosion, sedimentation, and rill or gully formation are much more likely to occur in, and downslope of, OHV-use areas located on or at the top of a slope than in/from a flat site or depressional area lacking watershed outlets. The GIS data layers could, therefore, identify areas to exclude from development for OHV use.

Although trail/area closures may be among the easiest management actions to implement, they may prove difficult to enforce, particularly under a policy of "closed unless posted open." The enforcement difficulty will depend on the number of trails involved, their locations relative to one another, and the number of enforcement personnel available. In the absence of funding for adequately monitoring regulation compliance, educating the public about the effects of their recreational pursuits may prove more economical and yield self-monitored trail users. In areas affected heavily by recreational activities, however, visitor management may be required (Jim, 1989). "Rationing" is a visitor management strategy that can be used to control the number of visitors over a given area where the available recreational resources are finite and/or unique. If the number of visitors is restricted, their quality of experience may be greater; a drawback of this approach is that the benefit is realized by fewer users (see Dimara and Skuras, 1998). Other strategies may include reserving a permit in advance, a permit lottery system, or implementing user fees that reflect the quality of experience (for example, it may cost more to use a high-use area that offers a high-quality experience than a low-use area that offers a low-quality experience). In many cases, these strategies could help alleviate pressures on law enforcement.

Wildlife is affected by OHV recreation in numerous ways, including displacement caused by human disturbance or direct mortality caused by vehicle-animal collisions (Cole and Landres, 1995; Knight and Cole, 1995; Miller, 1998; Stokowski and LaPointe, 2000; Cline and others, 2007). As OHV-related landscape fragmentation increases, habitat area and required juxtapositions of habitat types that meet the different needs of a given species, as well as adequate cover from disturbance, are diminished. Because the resulting effects may be detrimental to individuals and/or local populations (Knight and Cole, 1995), mitigation and management may be required to protect wildlife. One approach for mitigating the effects of wildlife displacement and disturbance may be to control the spatial and/or temporal proximity of OHV activities to wildlife, especially during critical nesting and breeding times (Gutzwiller, 1995). Again, GIS data layers could be very useful for identifying crucial wildlife areas and ensuring that they are not overlapped, fragmented, or otherwise disturbed by OHV-use areas.

4.2.3. Restoration of OHV-Impacted Areas

Revegetation of natural communities in arid environments is particularly difficult (Wallace and others, 1980) and studies evaluating revegetation have shown varying degrees of success (Graves and others, 1975, 1978; Kay and Graves, 1983; Grantz and others, 1998); thus, restoration of sites significantly degraded by OHV activity may be needed before sufficient levels of revegetation can take place, particularly if underlying bedrock has become exposed. Webb and others (1978) recommend that soil be imported and stabilized to replace the displaced soil where bedrock has become exposed. Generally, restoring soil horizons for re-establishing microbial communities can be achieved by inoculating soils with native microbes and mycorrhizae (Belnap, 1993; Bolling and Walker, 2000). Bolling and Walker (2000) suggest that decompacting OHV tracks and flattening out the lateral and center berms associated with them may increase the probability of community redevelopment with a more natural surface shape. Recovery of cryptobiotic soils, however, is more complex and may require long periods of time (Wilshire, 1 983b; Lovich and Bainbridge, 1999); ultimately, their recovery rate will depend on the degree of soil compaction and the nature and intensity of the initial disturbance (Bolling and Walker, 2000).

To reduce the potential for erosion at restoration sites, it is important to use mulch, stabilization techniques, and/or establish vegetation. Rasor (1976, cited *in* Webb and others, 1978) suggests that ground cover (such as wire netting) be applied across the restoration site to stabilize soils. Kay and Graves (1983) recommend that seeding with local seed stock begin as the disturbance desists. Revegetation techniques also may include container planting (Grantz and others, 1998), hand seeding (Lovich and Bainbridge, 1999), drill seeding (Kay, 1988), and establishment of visually dominant species, such as creosote bush (*Larrea tridentata*; Kay and Graves, 1983).

Chapter 5

MONITORING AND RESEARCH NEEDS

5.1. SUMMARY

More information is needed to help support policy making and land management as they pertain to the natural resources and people affected by OHV policies and management. Research needed to help support policy makers and land mangers includes (but is not limited to)

- well-designed studies that incorporate planned comparisons of treatment (OHV-impacted) and control (unimpacted/reference) sites, as well as "before and after OHV-impact" studies;
- studies at various spatial and temporal scales across all impacted habitat types;
- studies on habitat fragmentation and road-edge effects caused by OHV activities;
- studies on effects of various gradients in OHV disturbance and at varying distances from OHV routes;
- studies of OHV effects on plant and animal population dynamics;
- simultaneous evaluations of wildlife responses and OHV route-specific variables;
- studies to improve knowledge about the physical and chemical dynamics of soil compaction;
- studies evaluating the effects of erosion, sedimentation, and turbidity downslope of OHV-affected sites;
- improvements in techniques for successful site restoration;
- improvements in techniques and technologies for assessing OHV impacts over large areas and long time periods;
- studies that evaluate the effectiveness of various techniques to manage OHV use and its ecological and socioeconomic effects while simultaneously providing the greatest satisfaction among all land users; and
- studies that determine the economic and sociological costs of OHV use.

The experimental design of past studies on the ecological effects of OHVs often proved inadequate for providing reliable, defensible results. In particular, use of comparable treatment and control sites with adequate replication has been minimal. To better elucidate

OHV effects on wildlife, habitats, and vegetation, there is a need for well-replicated research based on treatments and controls ranging across various spatial and temporal scales within the full range of habitat types represented on BLM lands. In desert ecosystems, for example, the impacts of OHV activities can occur at several spatial and temporal scales (Forman and others, 2003: p. 129-134; Matchett and others, 2004; Brooks and Lair, 2005; see discussion in section 2.1); thus, research conducted across the scale(s) at which OHV activities and ecosystem responses are likely to occur will produce the most reliable information about OHV effects on land health and users of the land (Brooks and Lair, 2005). There also is a need for long-term monitoring of OHV effects at both designated OHV sites and undesignated (rogue use) sites, as well as revegetation sites.

Monitoring and research approaches that take advantage, and push the advancement, of existing and emerging technologies are needed to fully represent the scale and diversity of OHV impacts likely affecting plant and animal populations and communities, and to ascertain indicator thresholds as they pertain to BLM's land health standards. Current technologies, including satellite imagery, GPS, and GIS, among others, would be extremely useful in broadening the scope of OHV-impact research from site-based effects to ecosystems and landscapes. Technology also provides opportunities for better assessing OHV effects on wildlife, vegetation, and other natural resources. Finally, multiple assessments and the simultaneous recording of independent and dependent variables would improve overall results and better inform management decisions.

5.2. Monitoring and Research Needs

Remaining questions about effects of OHVs on ecosystems and people are numerous and varied. Information regarding management approaches for sustaining or restoring resources—from the level of single OHV routes to entire landscapes—to pre-disturbance conditions while still providing for quality OHV experiences is especially sparse. Therefore, the need for solid, well-designed research for supporting management decisions cannot be understated. Based on major unresolved issues and questions raised in the OHV impacts literature, current research needs include (but are not limited to)

- well-designed research capable of producing scientifically sound results by incorporating planned comparisons of treatment (OHV-impacted) and control (unimpacted/reference) sites, as well as studies that take advantage of opportunities to compare "before and after OHV-impacts" at sites that may be slated for—but are not yet impacted by—OHV use;
- studies to evaluate OHV effects on natural resource attributes at various spatial and temporal scales, particularly those appropriate for evaluating and understanding effects occurring at watershed, landscape, and plant and animal population levels;
- studies to improve the overall understanding of habitat fragmentation and road- edge effects caused by OHV activities in OHV-impacted habitat types;
- studies that evaluate effects of OHV activities at various gradients in OHV disturbance levels and at varying distances from OHV routes;

- studies to evaluate how OHV activities and habitat fragmentation affect plant and animal population dynamics;
- simultaneous evaluations of wildlife responses (from individual- to population- level scales) and route-specific variables, including route type and width, and the intensity, noise levels, and speed of traffic;
- studies to improve the understanding of the physical and chemical dynamics, as well as the consequences, of soil compaction, sedimentation, and turbidity downslope of OHV-affected sites;
- improvements in techniques and technologies for assessing OHV impacts over large areas and long periods of time;
- improvements in techniques for successful site restoration;
- studies to determine the economic value of ecosystem goods and services provided by natural resources on or in BLM (and affected) lands and waters; and
- studies to determine the costs of OHV use, including degradation or loss of ecosystem goods and services, loss of supportive constituencies, managing/enforcing regulations of OHV use, and restoring OHV-impacted sites.

5.2.1. Scientifically Rigorous Research Projects

Several studies discussed in this document compared areas impacted (treatment) to areas unimpacted (control) by OHVs. Many studies, however, did not compare treatment and control sites, or the control and treatment sites differed with regard to some major factor (for example, other recreation activities, livestock grazing, logging, or energy-development activities) that could have masked true differences pertaining to OHV effects. In other words, controls in research and monitoring programs provide the necessary frame of reference for identifying true effects of the variable(s) of interest; in turn, identifying true effects will better inform management actions. Therefore, among the greatest research needs pertaining to OHV effects on natural resources are scientifically defensible studies based on planned comparisons of OHVimpacted and unimpacted sites that are otherwise similar in terms of soils, topographies, climatic patterns, plant and animal communities, non-motorized activities, and other potentially confounding factors. This is particularly important for developing appropriate threshold indicator values for sustaining current resource conditions or triggering management actions.

An additional approach to research that can provide informative results is to conduct before-and-after comparisons of sites previously unimpacted by OHV use but slated for future OHV use. Although year-to-year variations in climate and plant or animal population cycles can introduce too much variation in ecological before-and-after data to provide statistically meaningful results, this research approach can provide valuable information when conditions are reasonably similar during each phase of the project. Thus, it would be prudent to take advantage of such opportunities as they come up.

Technologies and tools that could prove very useful in study design and site selection pertaining to OHV effects include satellite imaging and GIS. Although they are still advancing rapidly in terms of their capabilities and utilities, they are nonetheless already very helpful in studies ranging from the site to the landscape scale. Satellite imagery can help locate sites with differing extents of OHV activity, and, when rendered into GIS database layers, they can provide opportunities for repeating imagery over time and evaluating landscape-scale changes. U.S.

Geological Survey scientists, for example, are currently using these tools and technologies to study the effects of energy-development activities in sage-steppe systems on BLM lands in Wyoming (C. Aldridge, pers. comm.).

5.2.2. OHV Effects at Various Spatial and Temporal Scales, across Habitat Types

Past studies regarding OHV impacts on natural resources have focused primarily on effects at the single route or site level; thus, the overall understanding of landscape-, watershed-, and population-scale effects, including habitat/population fragmentation, is inadequate for managing OHV effects at larger spatial and longer temporal (Boyle and Samson, 1985) scales. Most of what is known pertains to very localized and short-term effects on sub-populations. Even then, past studies evaluating changes in animal densities of sub-populations may have violated basic assumptions of closed-population status. That is, studies that "failed to detect OHV effects on animal densities" may have been confounded by the immigration of new individuals after original individuals experienced poor survivorship due to direct or indirect effects of OHV activities. Mark-recapture studies that take advantage of radio-marking technologies, both in and away from the affected site, can help evaluate the extent to which assumptions of closed versus open populations are upheld or violated.

Research is also needed to better understand the edge effects of roads on different habitats and populations of different species or taxonomic groups. Here again, scale is important. Many past studies have evaluated the effects of traffic in immediately adjacent habitats (road rights-of-way, for example), whereas the effects may be realized well into the habitat interior quite far from travel routes. Thus, study designs that incorporate gradients of distance away from OHV routes or route networks would be crucial.

5.2.3. Research regarding Effects of OHVs on Animal Populations

Many prior studies of OHV impacts on wildlife have focused on indirect indicators of animal population health, such as distribution and behavior, at the expense of more direct indicators, such as population trend/size, gender and age ratios, and productivity. In part, this is because such data can be significantly more difficult and expensive to collect than behavioral and distribution data. Studies that address possible changes in plant and animal genetics in ecosystems fragmented by OHV roads and trails also are needed. For species that naturally occur at low densities, such as the desert tortoise, bighorn sheep (*Ovis canadensis*), and mountain lion, this is particularly important, as isolation of their populations could more easily lead to localized extinctions.

Understanding direct effects of OHV activities on wildlife populations would be further enhanced by research that evaluates wildlife population dynamics in different habitat types (Bury and others, 1977), at different spatial and temporal scales, and under different conditions (levels) of OHV use. Likewise, research programs that incorporate representative and disparate taxonomic groups would help identify ecosystem-level effects. For example, a given habitat- fragmentation factor (a network of OHV routes) might not have any significant impact on large ungulates, but it might lead to complete loss of genetic diversity among flightless invertebrates. In conjunction with this type of research, it would be important to identify which specific environmental and anthropogenic factors promote or limit the exchange of individuals across OHV-impacted habitat types or landscapes.

Few past studies have employed multiple-assessment or simultaneous survey methods to detect rare or sensitive species that may not be detected through standard monitoring techniques, resulting in biases towards more common and easily detected species and an incomplete understanding of the community at risk. For example, multiple survey techniques can be used to determine whether changes in species composition are due to changes in detectability across habitat types, times of day or season, geographical area, and abundance. For detecting reptiles and amphibians, researchers could combine noosing, pitfall sampling, night surveys, and road driving surveys. For mammal studies, track, scat, and camera surveys might enhance species detections. Many avian-ecology researchers now use combinations of road driving, point sampling, and mist netting, in addition to double-observer methods, for improving overall survey results.

Studies of OHV effects on wildlife also could be improved through simultaneous recording of independent and dependent variables. For example, many studies relating effects of OHVs on animal populations also lack any measurements of static and dynamic road- and OHVrelated variables—such as width and traffic intensity/noise levels (Andrews, 1990)—that may strongly influence species behavior, distribution, abundance, survivorship, and productivity. Given the importance of relationships between these independent and dependent variables, employing long-term monitoring that incorporates their simultaneous measurement would be very helpful (see Andrews, 1990). This type of information would be particularly useful for identifying indicator thresholds that might trigger area closures, re-routing of OHV routes, and/or implementing a restoration project. A variety of technologies and tools now available would be useful in these endeavors, including satellite telemetry, Global Positioning Systems (GPS), infrared photography, pneumatic vehicle counters, and so on.

Finally, there are a number of specific questions that researchers could ask with respect to OHV effects on animal populations. For example, the discussion in section 2.3.3 regarding edge effects of OHV routes raises the question of whether high densities of animals in roadside habitats represent favorable conditions for native fauna or dominance by invasive or non-native organisms and, if the former, whether these habitats are population sources or sinks (see Van Horne, 1983).

5.2.4. Research to Determine Socioeconomic Costs Associated with OHV Use

The literature search conducted for this report yielded no published studies of the economic costs associated with OHV use. Costs could include the degradation or loss of crucial ecosystem goods and services (such as a decline in water quality due to accelerated sedimentation and increased turbidity in wetlands caused by OHV traffic, or the loss of livestock and wildlife forage due to soil compaction caused by OHV traffic), the loss of economic and political support from both OHV users and non-motorized constituencies whose recreation experiences (or other land uses) are degraded by OHV effects, and the costs of managing OHV recreation and restoring sites impacted by OHV traffic. Although the costs of OHV effects may be challenging to assess, there are efforts underway to first identify the economic values of ecosystem goods and services and factor them into economic analyses of human activities. For example, Ducks Unlimited Canada and The Nature Conservancy have co-published a report that calls for the immediate acceleration of "efforts to measure, protect, and enhance the natural capital of Canada," including the need to "invest in science to measure, value, and monitor ecological goods and services, and develop economic instruments that recognize and protect natural capital, rather than continue to reward its destruction" (Olewiler, 2004). A companion report details the values of goods and services provided by wetlands (Gabor and others, 2004). Similar efforts to place a value on the goods and services provided by lands affected by OHV activities would provide the necessary basis for balancing economic equations of OHV use.

5.2.5. Research to Improve Site Restoration

The success of site restoration ultimately depends on the ability to return soils, including abiotic crusts and desert pavement, vegetation, and biotic crusts to their original condition. Therefore, research is needed to study the mechanisms behind, and mitigation that could diminish, changes in soils. Adams and others (1982) called for research on the disproportionate hardening of only slightly compacted desert soil and whether soil hardening is caused by chemical cementation or by a greater number of interstitial water bonds remaining between soil particles after drying. Webb and others (1978) called for studies that explore the amount of time required for soils to respond to revegetation of OHV sites and develop ways of mitigating changes in soil properties during OHV use. More recently, Bolling and Walker (2000) expressed the need for long-term monitoring and analysis of soil microbial populations. Answers to these and other important questions that elucidate the ways in which systems recover from the effects of OHVs are crucial to the possibility of long-term OHV use without incurring irreparable damage to ecosystems.

Chapter 6

CONCLUSION

It is apparent from the literature identified and discussed herein that the effects of OHV activities on ecosystems are diverse and potentially profound, if poorly understood. Studies have revealed a variety of effects on soil properties, watersheds, and vegetation resulting from one to multiple passes by OHV vehicles. Likewise, research has shown a variety of OHV effects on both OHV users and non-motorized recreators. Considerably less is known about impacts to wildlife or air and water quality. Whereas the results of past OHV-effects research have been reasonably consistent in demonstrating the nature of OHV effects in the immediate vicinity of single trails and OHV sites, there is a need for stronger emphasis on the cumulative effects— both spatial and temporal—of OHV use. For example, the effects of a single OHV route on a watershed may be greater when it is part of a route network than when it is the only route within the watershed. Furthermore, a network of OHV routes is likely to accommodate more users than a single route. Therefore, route density is an important consideration when evaluating and monitoring OHV effects across a given landscape.

The Council on Environmental Quality's (CEQ) regulations for implementing the National Environmental Policy Act define a cumulative impact as "...the impact on the environment that results from the incremental impact of the action when added to other past, present, and reasonably foreseeable future actions regardless of what agency (Federal or non-Federal) or person undertakes such other actions" (Council on Environmental Quality, 1997). In other words, cumulative impacts can result from what may appear on small spatial or temporal scales to be minor, but which become collectively significant when such actions take place over large areas and long periods of time. Moreover, one effect may interact with other effects to generate additional effects that are not apparent when evaluating effects individually.

The concept of cumulative impacts as they relate to OHV activity, therefore, must be applied in a landscape context, as these impacts are not site-specific and may affect adjacent or even more remote habitats and landscapes. For example, dust created from OHV activities can be dispersed to areas far away from habitats directly impacted by OHV activities. Likewise, erosion of soils during heavy rain events may increase sedimentation far downstream of areas directly subjected to OHV activities, and edge or corridor effects of OHV routes may promote widespread dispersal of non-native and invasive species. Thus, there is a need for greater monitoring and research emphasis on the effects of OHV activities not only in the areas directly subjected to those activities, but across impacted habitat types, watersheds, and landscapes. Overall, monitoring of cumulative impacts is needed at a scale larger than the physical imprint of the OHV-use area. By the same token, economic analyses of OHV use are needed to account for not only the immediate and apparent economic benefits, but also the long-term, large-scale, and ongoing costs associated with OHV use. Without factoring these variables into models of economic impacts, true cost:benefit ratios of OHV use will remain unknown.

Chapter 7

LITERATURE CITED

Absher, J. D., and Bright, A. D., 2004, Communication research in outdoor recreation and natural resources management, *in* Manfredo, M.J., Vaske, J.J., Bruyere, B.L., Field, D.R., and Brown, P.J., eds., *Society and Natural Resources—A summary of knowledge: Jefferso*n, Missouri, Modern Litho, p.1 17–1 26.

Achana, F., 2005, 2005 *ATV/motorbike user survey:* Boise, Idaho, Idaho Department of Parks and Recreation, Report, 37 p.

Adams, E.S., 1975, Effects of lead and hydrocarbons from snowmobile exhaust on brook trout (*Salvelinus fontanlis*): *Transactions of the American Fisheries Society,* v. 104, no. 2, p. 363–373.

Adams, J.A., Endo, A.S., Stolzy, L.H., Rowlands, P.G., and Johnson, H.B., 1982, Controlled experiments on soil compaction produced by off-road vehicles in the Mojave Desert, California: *Journal of Applied Ecology,* v. 19, no. 1, p. 167–175.

Adams, L.W., and Geis, A.D., 1983, Effects of roads on small mammals: *Journal of Applied Ecology,* v. 20, no. 2, p. 403–415.

American Motorcyclist Association, 1979, *The economic impact of off-highway motorcycling in southern California and Mexico:* Westerville, Ohio, Department of Government Relations, Technical Report, 14 p.

Amezketa Lizarraga, E., Gazol Lostao, R., and Aragues Lafarga, R., 2002, Development of an automated infiltrometer and its applicability to the field: *Investigacion Agraria Produccion y Proteccion Vegetales,* v. 17, no. 1, p. 13 1–142.

Anderson, D.H., Lime, D, W., and Wang, T.L., 1998, *Maintaining the quality of park resources and visitor experiences—A handbook for managers:* St. Paul, Minnesota, University of Minnesota Extension Tourism Center, TC-777, 149 p.

Andrews, A., 1990, *Fragmentation of habitat by roads and utility corridors—A review:* Australian Zoologist, v. 26, p. 130–141.

Angold, P.G., 1997, The impact of a road upon adjacent heathland vegetation—Effects on plants species composition: *Journal of Applied Ecology,* v. 34, no. 2, p. 409–417.

Arnold, J.L., and Koel, T.M., 2006, Effects of snowmobile emissions on the chemistry of snowmelt runoff in Yellowstone National Park—Final report: Yellowstone National Park, Wyoming, Report YCR-2006-1, Fisheries and Aquatic Sciences Section, Center for Resources, available on the Internet at *http://www.nps.gov/yell/planyourvisit /upload/snmbsnowmelt.pdf* (accessed April 2007).

Ashley, E.P., and Robinson, J.T., 1996, Road mortality of amphibians, reptiles, and other wildlife on the Long Point Causeway, Lake Erie, Ontario: *The Canadian Field-Naturalist,* v. 110, p. 403–412.

Bakowski, C., and Kozakiewicz, M., 1988, *The effect of forest road on bank vole and yellow-necked mouse populations: Acta Theriologica,* v. 33, no. 12–25, p. 345–353.

Barrett, H., Cagney, J., Clark, R., Fogg, J., Gebhardt, K., Hansen, P.L., Mitchell, B., Prichard, D., and Tippy, D., 1995, Riparian area management—*Process for assessing proper functioning condition: Denver, Colorado, U.S. Bureau of Land Management, Technical Reference* 1737-9, BLM/SC/ST-93/003 + 1737 +REV95.

Bashore, T.L., Tzilkowski, W.M., and Bellis, E.D., 1985, Analysis of deer-vehicle collision sites in Pennsylvania: *Journal of Wildlife Management,* v. 49, no. 3, p. 769–774.

Baur, A., and Baur, B., 1990, Are roads barriers to dispersal in the land snail *Arianta arbustorum*?: *Canadian Journal of Zoology,* v. 68, no. 33, p. 613–617.

Bazzaz, F.A., and Garbutt, K., 1988, The response of annuals in competitive neighborhoods— Effects of elevated CO2: Ecology, v. 69, no. 4, p. 937–946.

Beier, P., 1993, Determining minimum habitat areas and habitat corridors for cougars: *Conservation Biology,* v. 7, no. 11, p. 94–108.

Belnap, Jayne, 1993, Recovery rates of cryptobiotic crusts—Inoculant use and assessment methods: *Great Basin Naturalist,* v. 53, no. 1, p. 89–95.

Belnap, Jayne, 2002, Impacts of off-road vehicles on nitrogen cycles in biological soil crusts— Resistance in different U.S. deserts: *Journal of Arid Environments,* v. 52, no. 2, p. 155–165.

Belnap, Jayne, and Gardner, J.S., 1993, Soil microstructure in soils of the Colorado Plateau—The role of the cyanobacterium *Microcoleus vaginatus*: *Great Basin Naturalist,* v. 53, no. 1, p. 40–47.

Benson, P.E., Nokes, W.A., Cramer, R.L., O`Connor, J., Lindeman, W., 1986, Air-quality and noise issues in environmental planning: Washington, D.C., Transportation Research Board and National Research Council Technical Report, 72 p., *http://www.osti.gov/energycitations/product.biblio.jsp?osti_id=7172765.*

Berry, K.H., 1980a, A review of the effects of off-road vehicles on birds and other vertebrates, *in* DeGraaf, R.M., and Tilghman, N.G., eds., Management of western forests and grasslands for nongame birds—Workshop proceedings, Salt Lake City, Utah, February 11-14, 1980: Ogden Utah, *U.S. Forest Service, Intermountain Forest and Range Experiment Station, General Technical Report INT*–86, p. 451–467.

Berry, K.H., 1980b, The effects of four-wheel vehicles on biological resources, *in* Andrews, R.N.L., and Nowak, P., eds., Off-road vehicle use—A management challenge: Washington, D.C., U.S. *Office of Environmental Quality,* p. 231-233.

Berry, K.H., Shields, T., Woodman, A.P., Campbell, T., Roberson, J., Bohuski, K., and Karl, A., 1986, Changes in desert tortoise populations at the Desert Tortoise Research Natural Area between 1979 and 1985, *Desert Tortoise Council proceedings of symposium,* Palmdale, California, March 22-24, 1986: Desert Tortoise Council, p. 100–123.

Bight, A.D., Coredell, H.K., Hoover, A.P., and Tarranr, M.A., 2003, *A human dimensions framework—Guidelines for conducting social assessments:* Asheville, North Carolina, U.S. Forest Service, Southern Research Station, 83 p.

Boarman, W.I., Beigel, M.L., Goodlett, G.C., and Sazaki, M., 1998, *A passive integrated transponder system for tracking animal movements*: Wildlife Society Bulletin, v. 26, no. 4, p. 886–891.

Boarman, W.I., and Sazaki, M., 1996, Highway mortality in desert tortoises and small vertebrates—Success of barrier fences and culverts, *in* Evink, G.L., Garrett, P., Zeigler, D., and Berry, J., eds., *Trends in addressing transportation related wildlife mortality—Proceedings of the transportation related wildlife mortality seminar,* Orlando, Florida, April 30–May 2, 1996: Tallahassee, Florida, State Department of Transportation, Report no. FL-ER-58-96, *http://www.icoet.net/ICOWET/96proceedings. asp.*

Boarman, W.I., Sazaki, M., Berry, K.H., Goodlett, G.O., Jennings, W.B., and Woodman, A.P., 1993, Measuring the effectiveness of a tortoise-proof fence and culverts—Status report from first field season, *Proceedings of the Desert Tortoise Council: Desert Tortoise Council,* p. 126–142.

Boarman, W.I., Sazaki, M., and Jennings, W.B., 1997, The effects of roads, barrier fences, and culverts on desert tortoise populations in California, USA, *in* Abbeema, J.V., ed., *Proceedings—Conservation, restoration, and management of tortoises and turtles—An international conference,* State University of New York, Purchase, New York, July 1993: Orange, New Jersey, New York Turtle and Tortoise Society, p. 54–58.

Boer Leffe, W. J., 1958, *De entomologische waarde van eiken-berkenbos*: De Levendc Natuur, v. 61, p. 97-102.

Bolling, J.D., and Walker, L.R., 2000, Plant and soil recovery along a series of abandoned desert roads: *Journal of Arid Environments,* v. 46, no. 11, p. 1–24.

Bowles, A.E., 1995, *Responses of wildlife to noise, in* Knight, R.L., and Gutzwiller, K.J., eds., Wildlife and recreationists—Coexistence through management and research: Washington, D.C., Island Press, p. 109–156.

Boyle, S.A., and Samson, F.B., 1985, Effects of nonconsumptive recreation on wildlife—A review: *Wildlife Society Bulletin,* v. 13, p. 110–116.

Brattstrom, B.H., and Bondello, M.C., 1983, Effects of off-road vehicle noise on desert vertebrates, *in* Webb, R.H., and Wilshire, H.G., eds., *Environmental effects of off-road vehicles—Impacts and management in arid regions:* New York, Springer-Verlag, p. 167–206.

Brody, A.J., and Pelton, M.R., 1989, Effects of roads on black bear movements in western North Carolina: *Wildlife Society Bulletin,* v. 17, p. 5–10.

Brooks, J.J., and Champ, P.A., 2006, Understanding the wicked nature of "unmanaged recreation" in Colorado's Front Range: Environmental Management, v. 38, no. 5, p. 784–798, *http://www.springerlink.com/content/ek62272u626u6074/fulltext.pdf.*

Brooks, M.L., 1995, Benefits of protective fencing to plant and rodent communities of the western Mojave Desert, California: *Environmental Management,* v. 19, no. 1, p. 65–74.

Brooks, M.L., 1999, Effects of protective fencing on birds, lizards, and black-tailed hares in the western Mojave Desert: *Environmental Management,* v. 23, no. 3, p. 387–400.

Brooks, M.L., and Lair, Bridget, 2005, Ecological effects of vehicular routes in a desert ecosystem: Henderson, Nevada, U.S. Geological Survey, Western Ecological Research Center, Las Vegas Field Station, Technical Report, 23 p., *http://www.dmg.gov/documents/Desert_Road_Ecology_report.pdf* (accessed April 4, 2006).

Bruinderink, G.W.T.A., and Hazebroek, E., 1986, Ungulate traffic collisions in Europe: *Conservation Biology,* v. 10, no. 4, p. 1059–1067.

Bury, R.B., 1980, What we know and do not know about off-road vehicle impacts on wildlife, *in* Andrews, R.N.L., and Nowak, P., eds., Off-road vehicle use—A management challenge: Washington, D.C., *U.S. Office of Environmental Quality,* p. 110–122.

Bury, R.B., and Luckenbach, R.A., 1986, *Abundance of desert tortoises (Gopherus agassizii) in natural and disturbed habitats:* Fort Collins, Colorado, U.S. Fish and Wildlife Service, unpublished draft report.

Bury, R.B., Luckenbach, R.A., and Busack, S.D., 1977, Effects of off-road vehicles on vertebrates in the California desert USA: Washington, D.C., U.S. Fish and Wildlife Service, *Wildlife Research Reports* no. 8, p. 1–23.

Bury, R.L., and Fillmore, E.R., 1974, *Design of motorcycle areas near campgrounds—Effects on riders and non riders*: College Station, Texas, Department of Recreation and Parks, Texas A & M University, Technical Report, 72 p.

Bury, R.L., Holland, S.M., McEwen, D.N., 1983, Analyzing recreational conflict—Understanding why conflict occurs is requisite to managing that conflict: *Journal of Soil and Water Conservation,* v. 3, no. 5, p. 401-403.

Busack, S.D., and Bury, R.B., 1974, Some effects of off-road vehicles and sheep grazing on lizard populations in the Mojave Desert: *Biological Conservation,* v. 6, no. 3, p. 179–183.

Byrd, D.S., Gilmore, J.T., and Lea, R.H., 1983, Effect of decreased use of lead in gasoline on the soil of a highway: *Environmental Science and Technology,* v. 17, p. 12 1–123.

Caldewell, T.G., McDonald, E.V., and Young, M.H., 2006, Soil disturbance and hydrologic response at the National Training Center, Fort Irwin, California: *Journal of Arid Environments,* v. 67, no. 3, p. 456-472.

Clark, W.D., and Karr, J.R., 1979, Effects of highways on red-winged blackbird and horned lark populations: *The Wilson Bulletin,* v. 91, p. 143–145.

Clevenger, A.P., 1998, Permeability of the Trans-Canada Highway to wildlife in Banff National Park—Importance of crossing structures and factors influencing their effectiveness, *in* Evink, G.L., Garrett, P., Zeigler, D., and Berry, J., eds., *Proceedings of the International Conference on Wildlife Ecology and Transportation, Fort Myers,* Florida, February 10–12, 1998: Tallahassee, Florida, Florida Department of Transportation, Environmental Management Office Report no. FL DOT FL-ER 69-98, p. 109–119.

Cline, R., Sexton, Natalie, and Stewart, S.C., 2007, *A human-dimensions review of human-wildlife disturbance—A literature review of impacts, frameworks, and management solutions*: Fort Collins, Colorado, U.S. Geological Survey, Open-File Report 2007-1111, 91 p.

Cole, D.N., 1990, Trampling disturbance and recovery of cryptogamic soil crusts in Grand Canyon National Park: *Great Basin Naturalist,* v. 50, no. 4, p. 32 1–325.

Cole, D.N., 2004, Environmental impacts of outdoor recreation in wildlands, *in* Manfredo, M.J., Vaske, J.J., Bruyere, B.L., Field, D.R., and Brown, P.J., eds., *Society and natural resources—A summary of knowledge*: Jefferson, Missouri, Modern Litho, p. 107–116.

Cole, D.N. and Landres, P.B., 1995, Indirect effects of recreation on wildlife, *in* Knight, R.L., and Gutzwiller, K.J., eds., *Wildlife and recreationists—Coexistence through management and research:* Washington, D.C., Island Press, p. 183–202.

Committee on Rangeland Classification, Board on Agriculture, and National Research Council, 1994, *Rangeland Health—New methods to classify, inventory, and monitor rangelands:* Washington, D.C., National Academy Press.

Cordell, H.K., Betz, C.J., Green, G., and Owens, M., 2005, *Off-highway vehicle recreation in the United States, regions, and states—A national report from the national survey on recreation and the environment* (NSRE): U.S. Forest Service, Southern Research Station, Technical Report, 90 p., *http://www.srs.fs.usda.gov/pubs/21307.*

Cotts, N.R., Redente, E.F., and Schiller, R., 1991, Restoration methods for abandoned roads at lower elevations in Grand Teton National Park, *Wyoming: Arid Soil Research and Rehabilitation*, v. 5, p. 235–249.

Council on Environmental Quality, 1997, *Considering cumulative effects under the National Environmental Policy Act: Washington,* D.C., Council of Environmental Quality, Technical Report, *http://www.nepa.gov/nepa/ccenepa/ccenepa.htm.*

Crimmins, T., 1999, *Colorado off-highway vehicle user survey—Summary of results:* Denver, Colorado, Colorado State Parks, Technical Report, *http://www.outdoorlink.com/amtrails/resources/motors/motCoOHVsurvey/html.*

Daines, R.H., Motto, H., and Chilko, D.M., 1970, Atmospheric lead—Its relationship to traffic volume and proximity to highways: *Environmental Science and Technology,* v. 4, no. 4, p. 3 18–322.

Dave Miller Associates, 1981, *An economic/social assessment of snowmobiling in Maine:* Windham, Maine, Dave Miller Associates, Technical Report, 52 p.

Davidson, E.D., and Fox, M., 1974, *Effects of off-road motorcycle activity on Mojave Desert vegetation and soil:* Madroño, v. 22, no. 8, p. 381–390.

Dean Runyan Associates, 2000, *Campers in California travel patterns and economic impacts:* Portland, Oregon, Dean Runyan Associates, Technical Report, 76 p.

Decker, D.J., Krueger, R.A., Bauer, Jr., R.A., Knuth, B.A., and Richmond, M.E., 1996, *From clients to stakeholders—A philosophical shift for fish and wildlife management: Human Dimensions of Wildlife,* v. 1, no. 1, p. 70-82.

Dhindsa, M.S., Sandhu, J.S., Sandhu, P.S., and Toor, H.S., 1988, Roadside birds in Punjab (India)—Relation to mortality from vehicles: *Environmental Conservation,* v. 15, no. 4, p. 303–310.

Dillman, D.A., 2007, *Mail and telephone surveys—The tailored design method*: New York, John Wiley & Sons, 375 p.

Dimara, E., and Skuras, D., 1998, Rationing preferences and spending behavior of visitors to a scarce recreational resource with limited carrying capacity: *Land Economics,* v. 74, no. 3, p. 317-327.

Dobson, A., Ralls, K., Foster, M., Soulé, M.E., Simberloff, D., Doak, D., Estes, J.A., Mills, L.S., Mattson, D., Dizro, R., Arita, H., Ryan, S., Norse, E.A., Noss, R.F., and Johns, D., 1999, Connectivity—Maintaining flows in fragmented landscapes, *in* Soulé, M.E., and Terborgh, J., eds., *Continental Conservation—Scientific foundations of regional reserve networks:* Washington, D.C., Island Press, p. 129–170.

Gabor, T.S., North, A.K., Ross, L.C.M., Anderson, J.S., and Raven, M., 2004, *Natural values— The importance of wetlands and upland conservation practices in watershed management:* Stonewall, Manitoba, Ducks Unlimited Canada. Available online at *http://www.ducks.ca/conserve/wetland_values/pdf/nvalue.pdf.*

Eckert, R.E.J., Wood, M.K., Blackburn, W.H., and Peterson, F.F., 1979, Impacts of off-road vehicles on infiltration and sediment production of two desert soils: *Journal of Range Management,* v. 32, no. 5, p. 394–397.

English, B., Menard, J., and Jensen, K., 2001, *Estimated economic impacts of off-highway vehicle special events:* Knoxville, Tennessee, Agri-Industry Modeling and Analysis Group, Department of Agricultural Economics, University of Tennessee, *http://aimag.ag.utk.edu/pubs/ohveventimpacts.pdf* (accessed May, 2007).

Evink, G.L., Garrett, P., Zeigler, D., and Berry, J., 1996, Trends in addressing transportation related wildlife mortality—*Proceedings of the transportation related wildlife mortality seminar,* Orlando, Florida, April 30–May 2, 1996: Tallahassee, Florida, State Department of Transportation, Report no. FL-ER-58-96, *http://www.icoet.net/ ICOWET/96proceedings. asp.*

Evink, G.L., Garrett, P., Zeigler, D., and Berry, J., eds., 1998, *Proceedings of the International Conference on Wildlife Ecology and Transportation,* Fort Myers, Florida, February 10–12, 1998: Tallahassee, Florida, Florida Department of Transportation, Environmental Management Office Report no. FL DOT FL-ER 69-98.

Ewert, A., and Shultis, J., 1999, Technology and backcountry recreation—Boon to recreation or bust for management?: *Technology and Leisure,* v. 70, no. 8, p. 23–31.

Falkengren-Grerup, U., 1986, *Soil acidification and its impact on ground vegetation: Oecologia,* v. 70, p. 339–347.

Feldhammer, G.A., Gates, J.E., Harman, D.M., Loranger, A.J., and Dixon, K.R., 1986, Effects of interstate highway fencing on white-tailed deer activity: *Journal of Wildlife Management,* v. 50, no. 497–503.

Ferris, R., and Taylor, G., 1995, Contrasting effects of elevated CO2 and water deficit on two native herbs: *New Phytologist,* v. 131, no. 4, p. 491–501.

Fisher, A.L., Blahna, D.J., and Bahr, R., 2001, *Off-highway vehicle uses and owner preferences in Utah: Logan, Utah, Institute for Outdoor Recreation and Tourism,* Department of Forest Resources, Utah State University, Report no. IORT PR2001–02, 80 p.

Flint, A.L., and Childs, S., 1984, Development and calibration of an irregular hole bulk density sampler: *Soil Science Society of America Journal,* v. 48, no. 2, p. 374–378.

Folz, R.B., 2006, Erosion from all terrain vehicle (ATV) trails on National Forest lands (abs.): St. Joseph, Michigan, *Proceedings of the 2006 American Society of Agricultural and Biological Engineers, http://asae.frymulti.com/abstract.asp?aid=21056&t=2.*

Forman, R.T.T., and Alexander, L.E., 1998, Roads and their major ecological effects: *Annual Review of Ecology and Systematics,* v. 29, p. 207–231.

Forman, R.T.T., Sperling, D., Bissonette, J.A., Clevenger, A.P., Cutshall, C.D., Dale, V.H., Fahrig, L., France, R., Goldman, C.R., Heanue, K., and others, 2003, *Road ecology—Science and solutions:* Washington, D.C., Island Press, 481 p.

Foster, M.L., and Humphrey, S.R., 1995, *Use of highway underpasses by Florida panthers and other wildlife:* Wildlife Society Bulletin, v. 23, no. 1, p. 95–100.

Fox, D.G., 1986, *Establishing a baseline/protocols for measuring air quality effects in wilderness:* U.S. Forest Service, Rocky Mountain Forest and Range Experimental Station, Technical Report, 85–91 p.

Froehlich, H.A., Miles, D.W.R., and Robbins, R.W., 1985, Soil bulk density recovery on compacted skid trails in central Idaho: *Soil Science Society of America Journal,* v. 49, p. 1015–1017.

Furniss, M.J., Flanagan, S.A., and McFadin, B.A., 2000, Hydrologically connected roads—An indicator of the influence of roads on chronic sedimentation, surface water hydrology, and exposure to toxic chemicals: U.S. Forest Service, *Stream Systems Technology Center,* Rocky Mountain Research Station, Technical Report, 4 p., *http://www.stream.fs.fed.us/streamnt/jul100/jul100_2.htm.*

Garland, T., and Bradley, W.G., 1984, Effects of a highway on Mojave Desert rodent populations: *American Midland Naturalist,* v. 111, p. 47–56.

Gibbs, J.P., 1998, Amphibian movements in response to forest edges, roads, and streambeds in southern New England: *Journal of Wildlife Management,* v. 62, no. 2, p. 584–589.

Gibeau, M.L., and Herrero, S., 1998, Roads, rails, and grizzly bears in the Bow River Valley, Alberta, *in* Evink, G.L., Garrett, P., Zeigler, D., and Berry, J., eds., *Proceedings of the International Conference on Wildlife Ecology and Transportation,* Fort Myers, Florida, February 10–12, 1998: Tallahassee, Florida, *Florida Department of Transportation, Environmental Management Office Report no.* FL DOT FL-ER 69-98, p. 104–108.

Gilbert, T., and Wooding, J., 1996, An overview of black bear roadkills in Florida 1976–1995, *in* Evink, G.L., Garrett, P., Zeigler, D., and Berry, J., eds., *Trends in addressing transportation related wildlife mortality—Proceedings of the transportation related wildlife mortality seminar,* Orlando, Florida, April 30–May 2, 1996: Tallahassee, Florida, State Department of Transportation, Report no. FL-ER-58-96, *http://www. icoet. net/ICO WET/96proceedings. asp.*

Gill, T.E., 1996, Eolian sediments generated by anthropogenic disturbances of playas—Human impacts on the geomorphic system and geomorphic impacts on the human system: *Geomorphology,* v. 17, no. 1–3, p. 207–228.

Gish, C.D., and Christensen, R.E., 1973, Cadmium, nickel, lead, and zinc in earthworms from roadside soil: *Environmental Science and Technology,* v. 7, p. 1060–1062.

Graefe, A.R., and Thapa, B., 2004, Conflict in natural resource recreation, *in* Manfredo, M.J., Vaske, J.J., Bruyere, B.L., Field, D.R., and Brown, P.J., eds., *Society and natural resources—A summary of knowledge*: Jefferson, Missouri, Modern Litho, p. 209–224.

Grant, T.J., 2005, Flat-tailed horned lizards (*Phrynosoma mcallii*)*—Population size estimation, effects of off-highway vehicles, and natural history:* Fort Collins, Colorado, M.S. thesis, Department of Wildlife Biology, Colorado State University, 84 p., *http://www.warnercnr.colostate.edu/~doherty/Tyler%20Grant%20M.S.%20Thesis_FINAL.p df.*

Grantz, D.A., Vaughn, D.L., Farber, R., Kim, B., Zeldin, M., Vancuren, T., and Campbell, R., 1998, Seeding native plants to restore desert farmland and mitigate fugitive dust and PM10: *Journal of Environmental Quality,* v. 27, p. 1209–1218.

Graves, W.L., 1978, Revegetation of disturbed sites in the Mojave Desert with native shrubs: *California Agriculture,* p. 4–5.

Graves, W.L., Kay, B.L., and Williams, W.A., 1975, Seed treatment of Mojave Desert shrubs: *Agronomy Journal,* v. 67, p. 773–777.

Griffiths, M., and Van Schaik, C.P., 1993, The impact of human traffic on the abundance and activity periods of Sumatran rain forest wildlife: *Conservation Biology,* v. 7, no. 3, p. 623–626.

Grimes, D.W., Miller, R.J., and Wiley, P.L., 1975, Cotton and corn root development in two field soils of different strength characteristics: *Agronomy Journal,* v. 67, p. 5 19–523.

Grimes, D.W., Sheesley, W.R., and Wiley, P.L., 1978, Alfalfa root development and shoot regrowth in compact soil of wheel traffic patterns: *Agronomy Journal,* v. 70, p. 955–958.

Groom, J., McKinney, T., Gant, T., and Winchell, C., 2005, Preliminary 2006 *Astragalus magdalenae* var. *peirsonii* survivorship and demography findings: Carlsbad, California, U.S. Fish and Wildlife Service Field Office, *http://www.fws.gov/carlsbad/Rules/PMV/USFWS%202006%20preliminary%20findings.pdf.*

Grue, C.E., O'Shea, T.J., and Hoffman, D.J., 1984, *Lead concentrations and reproduction in highway-nesting barn swallows* [*Hirundo rustica*]: Condor, v. 86, no. 4, p. 383–389.

Gutzwiller, K.J., 1995, Recreational disturbances and wildlife communities *in* Knight, R.L., and Gutzwiller, K.J., eds., *Wildlife and recreationists—Coexistence through management and research:* Washington, D.C., Island Press, p. 169–181.

Hanski, I., 1999, Metapopulation ecology: New York, Oxford University Press, 324 p.

Hanski, I., and Simberloff, D., 1997, The metapopulation approach, its history, conceptual domain, and application to conservation, *in* Hanski, L., Gilpin, M., and Hanski, I., eds., *Metapopulation biology—Ecology, genetics, and evolution:* San Diego, California, Academic Press, p. 5–26.

Harris, L.D., and Gallagher, P.B., 1989, New initiatives for wildlife conservation—The need for movement corridors, *in* Mackintosh, G., ed., *Preserving communities and corridors:* Washington, D.C., Defenders of Wildlife, p. 11–34.

Haupt, H.F., 1959, Road and slope characteristics affecting sediment movement from logging roads: *Journal of Forestry,* v. 57, no. 5, p. 329–339.

Havlick, D.G., 2002, *No place distance—Roads and motorized recreation on America's public lands:* Washington, D.C., Island Press, 297 p.

Hillel, D., and Tadmor, N., 1962, *Water regime and vegetation in central Negev Highlands of Israel: Ecology,* v. 43, no. 1, p. 33–41.

Hinckley, B.S., Iverson, R.M., and Hallet, B., 1983, Accelerated water erosion in ORV off-road vehicle-use areas, *in* Webb, R.H., and Wilshire, H.G., eds., *Environmental effects of off-road vehicles—Impacts and management in arid regions:* New York, Springer-Verlag, p. 8 1–96.

Holzapfel, C., and Schmidt, W., 1990, Roadside vegetation along transects in the Judean Desert: *Israel Journal of Botany,* v. 39, p. 263–270.

Hooks, C.L., and Jansen, I.J., 1986, Recording cone penetrometer developed in reclamation research: *Soil Science Society of America Journal,* v. 50, no. 1, p. 10–12.

Hudson, N.W., 1971, Soil conservation: Ithaca, New York, Cornell University Press, 320 p.

Huey, L.M., 1941, Mammalian invasion via the highway: *Journal of Mammalogy,* v. 22, p. 383–385.

Hunt, R., Hand, D.W., Hannah, M.A., and Neal, A.M., 1991, Response to CO2 enrichment in 27 herbaceous species: *Functional Ecology,* v. 5, no. 3, p. 410–421.

Isensee, E., and Luth, H.G., 1992, *Continuous recording of soil density:* Landtechnik, v. 47; no. 9, p. 449–452.

Iverson, R.M., Hinckley, B.S., and Webb, R.M., 1981, Physical effects of vehicular disturbances on arid landscapes: *Science, v.* 212, no. 4497, p. 915–917.

Jackson, E.L., and Wong, R.A., 1982, Perceived conflict between urban cross-country skiers and snowmobilers in Alberta: *Journal of Leisure Research,* v. 14, no. 1, p. 47-62.

Jackson, S.D., and Griffin, C.R., 1998, Toward a practical strategy for mitigating highway impacts on wildlife, *in* Evink, G.L., Garrett, P., Zeigler, D., and Berry, J., eds., *Proceedings of the International Conference on Wildlife Ecology and Transportation,* Fort Myers, Florida, February 10–12, 1998: Tallahassee, Florida, Florida Department of Transportation, Environmental Management Office Report no. FL DOT FL-ER 69-98, p. 17–22.

Jacob, G. R., and Schreyer, R., 1980, Conflict in outdoor recreation—A theoretical perspective: *Journal of Leisure Research,* v. 12, no. 4, p. 368-3 80.

Jim, C., 1989, *Visitor management in recreation areas: Environmental Conservation,* v. 16, no. 1, p. 19–32.

Johansen, J.R., 1993, Cryptogamic crusts of semiarid and arid lands of North America: *Journal of Phycology,* v. 29, no. 22, p. 140–147.

Johansen, K., Coops, N.C., Gergel, S.E., and Stange, Y., 2007, Application of high spatial resolution satellite imagery for riparian and forest ecosystem classification: *Remote Sensing of Environment,* v. 110, no. 1, p. 29-44.

Johnson, H.B., Vasek, F.C., and Yonkers, T., 1975, *Productivity, diversity and stability relationships in Mojave Desert roadside vegetation: Bulletin of the Torrey Botanical Club,* v. 102, p. 106–115.

Kaseloo, P.A., and Tyson, K.O., 2004, *Synthesis of noise effects on wildlife populations: Office of Research and Technology Services, Federal Highway Administration,* Report no. FHWA–HEP–06–016, 67 p.

Kay, B.L., 1988, Artificial and natural revegetation of the second Los Angeles aqueduct: *Mojave Revegetation Notes,* no. 24, p. 1–3.

Kay, B.L., and Graves, W.L., 1983, History of revegetation studies in the California deserts, *in* Webb, R.H., and Wilshire, H.G., eds., *Environmental effects of off-road vehicles—Impacts and management in arid regions:* New York, Springer-Verlag, p. 315–324.

Keener, K.M., Wood, R.K., Holmes, R.G., and Morgan, M.T., 1991, Soil strength evaluation of sample cores in a field measurement system: St. Joseph, Michigan, *American Society of Agricultural Engineers*, ASAE Paper No. 91-1526, 10 p.

Kline, N.C., and Swann, D.E., 1998, Quantifying wildlife road mortality in Saguaro National Park, *in* Evink, G.L., Garrett, P., Zeigler, D., and Berry, J., eds., *Proceedings of the International Conference on Wildlife Ecology and Transportation,* Fort Myers, Florida, February 10–12, 1998: Tallahassee, Florida, Florida Department of Transportation, Environmental Management Office Report no. FL DOT FL-ER 69-98, p. 23–31.

Knight, R.L., and Cole, D.N., 1995, Wildlife response to recreationists, *in* Knight, R.L., and Gutzwiller, K.J., eds., *Wildlife and recreationists—Coexistence through management and research*: Washington, D.C., Island Press, p. 5 1–69.

Knight, R.L., and Kawashima, J.Y., 1993, Responses of raven and red-tailed hawk populations to linear right-of-ways: *Journal of Wildlife Management,* v. 57, no. 2, p. 265–271.

Komatsu, M., Kato, A., and Sakamoto, K., 1988, Study on the automatic measurement system of soil parameters. Development of soil hardness sensor: Tottori, Japan, Tottori University, *Bulletin of the Faculty of Agriculture*, v. 41, p. 39–46.

Kopperoinen, L., Shemeikka, P.J., and Lindblom, V., 2004, *Environmental GIS in the management of visitor flows:* Helsinki, Finland, Finnish Environment Institute, Working Papers of the Finnish Forest Research Institute 2, *http://www.metla.fi/julkaisut/workingpapers/2004/mwp002-58.pdf.*

Lacey, C.A., Lacey, J.R., Fay, P.K., Storey, J.M., and Zamora, D.L., 1997, *Controlling knapweed on Montana rangeland:* Bozeman, Montana, Montana State University Extension Service, Circular 311.

Lehnert, M.E., and Bissonette, J.A., 1997, Effectiveness of highway crosswalk structures at reducing deer-vehicle collisions: *Wildlife Society Bulletin,* v. 25, no. 4, p. 809–818.

Leung, Y.-F., and Meyer, K., 2004, Soil compaction as indicated by penetration resistance—A comparison of two types of penetrometers, *in* Harmon, D., Kilgore, B.M., and Vietzke, G.E. eds., Protecting our diverse heritage—The role of parks, protected areas, and cultural sites, *Proceedings of the 2003 George Wright Society* / National Park Service Joint Conference: Hancock, Michigan, The George Wright Society, *http://www.georgewright.org/2003proc.html* (accessed April 2007).

Liddle, M.J., 1997, *Recreation ecology—The ecological impact of outdoor recreation and ecotourism:* London, Chapman and Hall, 666 p.

Lightfoot, D.C., and Whitford, W.G., 1991, Productivity of creosotebush foliage and associated canopy arthropods along a desert roadside: *American Midland Naturalist*, v. 125, no. 2, p. 310–322.

Lindenmayer, D.B., Margules, C.R., and Botkin, D.B., 2000, Indicators of biodiversity for ecological sustainable management: *Conservation Biology,* v. 14, p. 941-950.

Lisle, T.E., and Eads, R.E., 1991, Methods to measure sedimentation of spawning gravels: Berkeley, California, U.S. Forest Service, *Pacific Southwest Research Station, Research* Note PSW-411, 7 p., *http://www.fs.fed.us/psw/rsl/projects/water/Lisle91.pdf* (accessed April 2007).

Lovallo, M.J., and Anderson, E.M., 1996, Bobcat (*Lynx rufus*) home range size and habitat use in northwest Wisconsin: *American Midland Naturalist*, v. 135, no. 2, p. 241–252.

Lovich, J.E., and Bainbridge, D., 1999, Anthropogenic degradation of the southern California desert ecosystem and prospects for natural recovery and restoration: *Environmental Management,* v. 24, no. 3, p. 309–326.

Lowery, B., and Morrison, J.E., Jr., 2002, Soil penetrometers and penetrability, *in* Dane, J., and Topp, C., eds., Methods of soil analysis, part 4—Physical methods: Madison, Wisconsin, *Soil Science Society of America,* p. 363–388.

Luckenbach, R.A., 1978, An analysis of off- road vehicle use on desert avifaunas: *Transactions of the North American Wildlife and Natural Resources Conference,* v. 43, p. 157–162.

Luckenbach, R.A., and Bury, R.B., 1983, Effects of off-road vehicles on the biota of the Algodones Dunes, Imperial County, California, USA: *Journal of Applied Ecology,* v. 20, no. 1, p. 265–286.

Luo, H.R., Smith, L.M., Allen, B.L., and Haukos, D.A., 1997, Effects of sedimentation on playa wetland volume: *Ecological Applications,* v. 7, p. 247–252.

Lyren, L., 2001, *Movement patterns of coyotes and bobcats relative to roads and underpasses in the Chino Hills area of southern California*: Pomona, California, California State Polytechnic University, Pomona, 127 p.

Mader, H.J., 1984, *Animal habitat isolation by roads and agricultural fields: Biological Conservation,* v. 29, no. 1, p. 8 1–96.

Major, M.J., 1987, *Managing off-the-road vehicles—A recurring round of events: Journal of Forestry,* v. 85, no. 11, p. 37-41.

Massachusetts Department of Environmental Management, 1995, *Report on policy for off-road vehicle use in Massachusetts forests and parks*: Boston, Massachusetts, Massachusetts Executive Office of Environmental Affairs.

Matchett, J.R., Gass, L., Brooks, M.L., Mathie, A.M., Vitales, R.D., Campagna, M.W., Miller, D.M., and Weigand, J.F., 2004, *Spatial and temporal patterns of off-highway vehicle use at the Dove Springs OHV Open Area,* California: Henderson, Nevada, U.S. Geological Survey, Western Ecological Research Center, 17 p., *http://www.werc.usgs.gov/lasvegas/pdfs/Matchett_et_al_2004_Spatial%20and%20Temporal%20Patterns%20of%20off-highway.pdf* (accessed May 2007).

McBrayer, M.C., Mauldon, M., Drumm, E.C., and Wilson, G.V., 1997, Infiltration tests on fractured compacted clay: *Journal of Geotechnical and Geoenvironmental Engineering,* v. 123, no. 5, p. 469–473.

McCaffery, K.R., 1973, Road-kills show trends in Wisconsin deer populations: *Journal of Wildlife Management,* v. 37, no. 2, p. 212–216.

Means, D.B., 1999, The effects of highway mortality on four species of amphibians at a small, temporary pond in northern Florida, *in* Evink, G.L., Garrett, P., and Zeigler, D., eds., *Proceedings of the third international conference on wildlife ecology and transportation,* Missoula, Montana, September 13–16: Tallahassee, Florida, Florida Department of Transportation, Report no. FL-ER-73-99, p. 125–128.

Meffe, G.K., and Carroll, R.C., 1997, *Principles of conservation biology: Sunderland, Massachusetts,* Sinauer Associates, Inc., 729 p.

Meine, C, 1998, Outdoor recreation in the United States—The quiet explosion, *in* Outdoor recreation—Promise and peril in the new West: Boulder, Colorado, *19th Annual Summer Conference,* Natural Resources Law Center, University of Colorado School of Law and the U.S. Bureau of Land Management.

Miller, R.E., Hazard, J., and Howes, S., 2001, *Precision, accuracy, and efficiency of four tools for measuring soil bulk density or strength:* U.S. Forest Service, Pacific Northwest Research Station, Report no. PNW–RP–532, 16 p.

Miller, S., 1998, Environmental impacts—The dark side of outdoor recreation, *in* Outdoor recreation—Promise and peril in the New West: Boulder, Colorado, *19th Annual Summer Conference,* Natural Resources Law Center, University of Colorado School of Law and U.S. Bureau of Land Management.

Mooney, H.A., Kueppers, M., Koch, G., Gorham, J., Chu, C., and Winner, W.E., 1988, *Compensating effects to growth of carbon partitioning changes in response to SO2-induced photosynthetic reduction in radish:* Oecologia, v. 75, no. 4, p. 502–506.

Moore, T.G., and Mangel, M., 1996, Traffic related mortality and the effects on local populations of barn owls *Tyto alba, in* Evink, G.L., Garrett, P., Zeigler, D., and Barry, J., eds., Trends in addressing transportation related wildlife mortality—*Proceedings of the transportation related wildlife mortality seminar,* Orlando, Florida, April 30–May 2, 1996: Tallahassee, Florida, State Department of Transportation, Report no. FL-ER-58-96, *http://www. icoet. net/ICO WET/96proceedings. asp.*

Motto, H.L., Daines, R.H., Chilko, D.M., and Motto, C.K., 1970, Lead in soils and plants—Its relationship to traffic volume and proximity to highways: *Environmental Science and Technology,* v. 4, p. 231–237.

Mumme, R.L., Schoech, S.J., Woolfenden, G.W., and Fitzpatrick, J.W., 2000, Life and death in the fast lane—*Demographic consequences of road mortality in the Florida scrub-jay: Conservation Biology,* v. 14, no. 2, p. 501–512.

Munguira, M.L., and Thomas, J.A., 1992, Use of road verges by butterfly and burnet populations, and the effect of roads on adult dispersal and mortality: *Journal of Applied Ecology,* v. 29, p. 316–329.

Nakata, J.K., 1983, Off-road vehicular destabilization of hill slopes—The major contributing factor to destructive debris flows in Ogden, Utah, 1979, *in* Webb, R.H., and Wilshire, H.G., eds., *Environmental effects of off-road vehicles—Impacts and management in arid regions:* New York, Springer-Verlag, p. 343–353.

Nakata, J.K., Wilshire, H.G., and Barnes, G.C., 1976, Origin of Mojave Desert dust plume photographed from space*:* *Geology,* v. 4, p. 644–648.

Nelson, C.M., and Lynch, J.A., 2001, *A usable pilot off-road vehicle project evaluation: East Lansing,* Michigan, Department of Park, Recreation and Tourism Resources, Michigan State University, Technical Report, 50 p.

Nelson, C.M., Lynch, J.A., and Stynes, D.J., 2000, Michigan licensed off-road vehicle use and users 1998–99: *East Lansing, Michigan, Department of Park, Recreation and Tourism Resources,* Michigan State University, Technical Report, 49 p.

Nicholson, L., 1978, *The effects of roads on desert tortoise populations, in* Trotter, M. (ed.), Proceedings of the 1978 Desert Tortoise Council Symposium, San Diego, California: Desert Tortoise Council, p. 127–129.

Noon, B.R., 2003, Conceptual models in monitoring ecological resources, *in* Busch, D.E., and Trexler, J.C., eds., *Monitoring ecosystems—Interdisciplinary approaches for evaluating ecoregional initiatives:* Washington, D.C., Island Press, p. 27-71.

Olewiler, N., 2004, The value of natural capital in settled areas of Canada: Stonewall, Manitoba, Ducks Unlimited Canada and the Nature Conservancy of Canada, 36 p. Available online at *http://www.ducks.ca/aboutduc/news/archives/pdf/ncapital.pdf.*

O'Sullivan, M.F., and Ball, B.C., 1982, A comparison of five instruments for measuring soil strength in cultivated and uncultivated cereal seedbeds: *Journal of Soil Science,* v. 33, no. 4, p. 597–608.

Ouren, D.S., and Watts, R.D., 2005, Public access management as an adaptive wildlife management tool: Fort Collins, Colorado, U.S. Geological Survey, *Open-File Report* 2005-1349, 10 p.

Oxley, D.J., Fenton, M.B., and Carmody, G.R., 1974, Effects of roads on populations of small mammals: *Journal of Applied Ecology,* v. 11, no. 1, p. 51–59.

Paquet, P., and Callahan, C., 1996, Effects of linear developments of winter movement of gray wolves in the Bow River Valley of Banff National Park, Alberta, *in* Evink, G.L., Garrett, P., Zeigler, D., and Berry, J., eds., *Trends in addressing transportation related wildlife mortality—Proceedings of the transportation related wildlife mortality seminar,* Orlando, Florida, April 30–May 2, 1996: Tallahassee, Florida, State Department of Transportation, Report no. FL-ER-58-96, *http://www.icoet.net/ICOWET/96proceedings.asp.*

Parendes, L.A., and Jones, J.A., 2000, *Role of light availability and dispersal in exotic plant invasion along roads and streams in the H. J. Andrews Experimental Forest,* Oregon: Conservation Biology, v. 14, no. 1, p. 64–75.

Pellant, M., Shaver, P., Pyke, D.A., and Herrick, J.E., 2005, *Interpreting indicators of rangeland health, version 4—Technical Reference 1734-6:* Denver, Colorado, U.S. Bureau of Land Management, National Science and Technology Center, Report no. BLM/WO/ST00/001+1734/REV05, 122 pp.

Petersen, T.R., ed., 2006, *A road runs through it—Reviving wild places:* Boulder, Colorado, Johnson Books, 240 p.

Phillips, A.M., III, and Kennedy, D.J., 2006, *Seed bank and survival of Peirson's milkvetch (Astragalus magdalenae* var. *peirsonii) in the Algodones Dunes,* California, 2005-06, final report prepared for the American Sand Association: Eckert, Colorado, Botanical and Environmental Consulting, 22 p.

Powers, R.F., Tiarks, A.E., and Boyle, J.R., 1998, *Assessing soil quality—Practicable standards for sustainable forest productivity in the United States, in The contribution of soil science to the development of and implementation of criteria and indicators of sustainable forest management:* St. Louis, Missouri, Soil Science Society of America, Inc., p. 53–80.

Prose, D.V., 1985, Persisting effects of armored military maneuvers on some soils of the Mojave Desert: *Environmental Geology and Water Sciences*, v. 7, no. 3, p. 163–170.

Prose, D.V., Metzger, S.K., and Wilshire, H.G., 1987, Effects of substrate disturbance on secondary plant succession—Mojave Desert, California: *Journal of Applied Ecology,* v. 24, no. 1, p. 305–313.

Quarles, H.D., Hanawalt, R.B., and Odum, W.E., 1974, Lead in small mammals, plants, and soil at varying distances from a highway: *Journal of Applied Ecology,* v. 11, no. 3, p. 937–949.

Rasor, R., 1976, Fair share: *American Motorcycle Association News,* August, p. 16–17.

Reed, P., and Haas, G., 1989, *Off highway vehicles in Colorado: estimated recreational use and expenditures:* Fort Collins, Colorado, Colorado State University, Department of Recreational Resource and Landscape Architecture, 13 p.

Reed, R.A., Johnson-Barnard, J., and Baker, W.L., 1996, Contribution of roads to forest fragmentation in the Rocky Mountains: *Conservation Biology,* v. 10, no. 4, p. 1098–1106.

Reijnen, R., Foppen, R., Braak, C.T., and Thissen, J., 1995, The effects of car traffic on breeding bird populations in woodland. III. Reduction of density in relation to the proximity of main roads: *Journal of Applied Ecology*, v. 32, no. 1, p. 187–202.

Reijnen, R., Foppen, R., and Meeuwsen, H., 1996, The effects of traffic on the density of breeding birds in Dutch agricultural grasslands: *Biological Conservation,* v. 75, p. 255–260.

Reijnen, R., Foppen, R., and Veenbaas, G., 1997, Disturbance by traffic of breeding birds—Evaluation of the effect and considerations in planning and managing road corridors: *Biodiversity and Conservation,* v. 6, no. 4, p. 567–581.

Ringold, P.L., Mulder, Barry, Alegria, Jim, Czaplewski, R.L., Tolle, Tim, and Burnett, Kim, 2003, Design of an ecological monitoring strategy for the forest plan in the Pacific Northwest, *in* Busch, D.E., and Trexler, J.C., eds., *Monitoring ecosystems—*

Interdisciplinary approaches for evaluating ecoregional initiatives: Washington, D.C., Island Press, p. 73-99.

Rocky Mountain Research Institute, 2002, *Off-road vehicles in Colorado—Facts, trends, recommendations*: Nederland, Colorado, Rocky Mountain Research Institute.

Romin, L.A., and Bissonette, J.A., 1996, *Temporal and spatial distribution of highway mortality of mule deer on newly constructed roads at Jordanelle Reservoir,* Utah: Great Basin Naturalist, v. 56, no. 1, p. 1–11.

Rosen, P.C., and Lowe, C.H., 1994, Highway mortality of snakes in the Sonoran Desert of southern Arizona: *Biological Conservation,* v. 68, no. 22, p. 143–148.

Rudolph, D.C., Burgdorf, S.J., Conner, R.N., and Dickson, J.G., 1998, *The impact of roads on the timber rattlesnake,* (*Crotalus horridus*), in eastern Texas, *in* Evink, G.L., Garrett, P., Zeigler, D., and Berry, J., eds., *Proceedings of the International Conference on Wildlife Ecology and Transportation,* Fort Myers, Florida, February 10–12, 1998: Tallahassee, Florida, Florida Department of Transportation, Environmental Management Office Report no. FL DOT FL-ER 69-98, p. 236–240.

Ruediger, B., 1998, Rare carnivores and highways—Moving into the 21st century, *in* Evink, G.L., Garrett, P., Zeigler, D., and Berry, J., eds., *Proceedings of the International Conference on Wildlife Ecology and Transportation,* Fort Myers, Florida, February 10–12, 1998:

Tallahassee, Florida, Florida Department of Transportation, *Environmental Management* Office Report no. FL DOT FL-ER 69-98, p. 10–16.

Samways, M.J., 1989, Insect conservation and landscape ecology—A case-history of bush crickets (Tettigoniidae) in southern France: *Environmental Conservation,* v. 16, no. 3, p. 217–226.

Sanders, T.G., and Addo, J.Q., 2000, Experimental road dust measurement device: *Journal of Transportation Engineering,* v. 126, no. 6, p. 530-535.

Schuett, M.A, and Ostergren, D., 2003, Environmental concern and involvement of individuals in selected voluntary associations: *Journal of Environmental Education,* v. 34, no. 4, p. 30–38.

Seibert, H.C., and Conover, J.H., 1991, Mortality of vertebrates and invertebrates on an Athens County, Ohio [USA], highway: Ohio *Journal of Science,* v. 91, no. 4, p. 163–166.

Servheen, C., Walker, J., and Kasworm, W., 1998, *Fragmentation effects of high-speed highways on grizzly bear populations shared between the United States and Canada, in* Evink, G.L., Garrett, P., Zeigler, D., and Berry, J., eds., Proceedings of the International Conference on Wildlife Ecology and Transportation, Fort Myers, Florida, February 10–12, 1998: Tallahassee, Florida, Florida Department of Transportation, Environmental Management Office Report no. FL DOT FL-ER 69-98, p. 97–103.

Shindler, B., Brunson, M.W., and Cheek, K.A., 2004, Social acceptability in forest and range management, *in* Manfredo, M.J., Vaske, J.J., Bruyere, B.L., Field, D.R., and Brown, P.J., eds., *Society and natural resources—A summary of knowledge:* Jefferson, Missouri, Modern Litho, p. 146–157.

Singer, F.J., 1978, Behavior of mountain goats in relation to US Highway 2, Glacier Park, Montana: *Journal of Wildlife Management*, v. 42, p. 59 1–597.

Smith, W.H., 1976, Lead contamination of the roadside ecosystem: *Journal of the Air Pollution Control Association,* v. 26, p. 753–766.

Spellerberg, I.F., 1998, Ecological effects of roads and traffic—A literature review: *Global Ecology and Biogeography Letters,* v. 7, no. 5, p. 3 17–333.

Spellerberg, I.F., and Morrison, T., 1998, *The ecological effects of new roads—A literature review:* Wellington, New Zealand, New Zealand Department of Conservation, Technical Report, 55 p.

Spencer, H.J., and Port, G.R., 1988, Effects of roadside conditions on plants and insects: II. Soil conditions: *Journal of Applied Ecology,* v. 25, no. 22, p. 709–715.

Stefanov, W.L., Ramsey, M.S., and Christensen, P.R., 2001, *Mapping of fugitive dust generation, transport, and deposition in the Nogales, Arizona region using Enhanced Thematic Mapper Plus (ETM+) data:* American Geophysical Union, 2001 Spring Meeting, Boston, Massachusetts, May 29–June 2, 2001, abstract no. B32A-10, *http://adsabs.harvard.edu/abs/2001AGUSM...B32A10S* (accessed May 2007).

Stokowski, P.A., and LaPointe, C.B., 2000, Environmental and social effects of ATVs and ORVs—An annotated bibliography and research assessment: Burlington, Vermont, University of Vermont, School of Natural Resources, 32 p., *http://atfiles.org/files/pdf/ohvbibliogVT00.pdf.*

Sullivan, B.K., 1981, Distribution and relative abundance of snakes along a transect in California: *Journal of Herpetology,* v. 15, no. 2, p. 247–248.

Swihart, R.K., and Slade, N.A., 1984, Road crossing in *Sigmodon hispidus* and *Microtus ochrogaster*: *Journal of Mammalogy,* v. 65, no. 2, p. 357–360.

Tewes, M.E., and Blanton, D.R., 1998, *Potential impacts of international bridges on ocelots and jaguarundis along the Rio Grande wildlife corridor, in* Evink, G.L., Garrett, P., Zeigler, D.,

and Berry, J., eds., Proceedings of the International Conference on Wildlife Ecology and Transportation, Fort Myers, Florida, February 10–12, 1998: Tallahassee, Florida, Florida Department of Transportation, *Environmental Management Office Report no.* FL DOT FL-ER 69-98, p. 135–139.

Thompson, C.R., Kats, G., and Lennox, R.W., 1980, Effects of SO2 and/or NO2 on native plants of the Mojave Desert and Eastern Mojave-Colorado Desert: *Journal of the Air Pollution Control Association,* v. 30, no. 12, p. 1304–1309.

Thompson, C.R., Olszyk, D.M., Kats, G., Bytnerowicz, A., Dawson, P.J., and Wolf, J.W., 1984, Effects of ozone or sulfur dioxide on annual plants of the Mojave Desert: *Journal of the Air Pollution Control Association,* v. 34, no. 10, p. 1017–1022.

Trombulak, S.C., and Frissell, C.A., 2000, Review of ecological effects of roads on terrestrial and aquatic communities: *Conservation Biology,* v. 14, no. 1, p. 18–30.

Tyser, R.W., and Worley, C.A., 1992, Alien flora in grasslands adjacent to road and trail corridors in Glacier National Park, Montana (U.S.A.): *Conservation Biology,* v. 6, no. 2, p. 253-262.

U.S. Bureau of Land Management, 2001, *Rangeland health standards handbook H-4180-1, release 4-107:* U.S. Bureau of Land Management, 50 p., *http://www.blm.gov/nhp/efoia/wo/handbook/h4180-1.pdf* (accessed April 2007)

U. S. Bureau of Land Management, 2006, *Roads and trails terminology:* Denver, Colorado, Technical Note 422. U.S. Bureau of Land Management, Technical Notes BLM/WO/ST-06/006, 67 p.

U.S. Bureau of Land Management, 2007, *Notice of final action to adopt revisions to the Bureau of Land Management's procedures for managing the NEPA process,* chapter 11 of the Department of the Interior's manual part 516: Federal Register, vol. 72, no. 156, part II, p. 45504- 45542.

Udevitz, M.S., Howard, C.A., Robel, R.J., and Curnutte, B., 1980, Lead contamination in insects and birds near an interstate highway, Kansas: *Environmental Entomology,* v. 9, no. 1, p. 35–36.

van der Zande, A.N., Keurs, W.J., and van der Weijden, W.J., 1980, The impact of roads on the densities of four bird species in an open field habitat—Evidence of a long-distance effect: *Biological Conservation,* v. 18, no. 4, p. 299–321.

Van Dyke, F.G., Brocke, R.H., and Shaw, H.G., 1986, Use of road track counts as indices of mountain lion presence: *Journal of Wildlife Management,* v. 50, no. 1, p. 102–109.

Van Horne, Beatrice, 1983, Density as a misleading indicator of habitat quality: *Journal of Wildlife Management,* vol. 47, no. 4, p. 893–901.

Vancini, F.W., 1989, Policy and management considerations for off-road vehicles—*Environmental and social impacts:* Ithaca, New York, Cornell University, 19 p.

Vasek, F.C., Johnson, H.B., and Brum, G.D., 1975, *Effects of power transmission lines on vegetation of the Mojave Desert:* Madroño, v. 23, no. 1, p. 114–130.

Vaske, J. J., Donnelly, M. P., Wittmann, K., and Laidlaw, S., 1995, *Interpersonal versus social- values conflict:* Leisure Sciences, v. 17, no. 3, p. 205-222.

Vaske, J.J, Fulton, D.C., and Manfredo, M.J., 2001, Human dimensions considerations in wildlife management planning, *in* Decker, D.J., Brown, T.L., and Siemer, W.F., eds., *Human Dimensions of Wildlife Management in North America: Bethesda,* Maryland, The Wildlife Society, p. 91-108.

Veen, J., 1973, *De verstoring van weidevogelpopulaties: Stedeb. en Volkshuisv.*, 5v. 3, p. 16-26.

von Seckendorff Hoff, K., and Marlow, R., 1997, Highways and roads are population sinks for desert tortoises, *in* New York Turtle and Tortoise Society, *Proceedings—Conservation, restoration, and management of tortoises and turtles—An international conference:* Orange, New Jersey, New York Turtle and Tortoise Society, p. 482.

Vos, C.C., and Chardon, J.P., 1998, Effects of habitat fragmentation and road density on the distribution pattern of the moor frog *Rana arvalis*: *Journal of Applied Ecology,* v. 35, no. 1, p. 44–56.

Walker, D.A., and Everett, K.R., 1987, Road dust and its environmental impact on Alaskan taiga and tundra: *Arctic and Alpine Research,* v. 19, no. 4, p. 479–489.

Wallace, A., Romney, E.M., and Hunter, R.B., 1980, The challenge of a desert: revegetation of disturbed desert lands, *in* Woods, S.L., ed., *Great Basin naturalist memoirs—Soil-plantanimal relationships bearing on revegetation and land reclamation in Nevada Deserts:* Provo, Utah, Brigham Young University, p. 216–225.

Ward, A.L., 1982, Mule deer behavior in relation to fencing and underpass on Interstate 80 in Wyoming: *Transportation Research Record,* v. 859, no. 8–13.

Watson, A., Asp, C., Walsh, J. and Kulla, A., 1997, The contribution of research to managing conflict among national forest users: *Trends,* v. 34, no. 3, p. 29–3 5.

Webb, R.H., 1982, Off-road motorcycle effects on a desert soil: *Environmental Conservation,* v. 9, no. 3, p. 197–208.

Webb, R.H., 1983, Compaction of desert soils by off-road vehicles, *in* Webb, R.H., and Wilshire, H.G., eds., *Environmental effects of off-road vehicles—Impacts and management in arid regions:* New York, Springer-Verlag, p. 50–79.

Webb, R.H., 2002, Recovery of Severely Compacted Soils in the Mojave Desert, California, USA.: *Arid Land Research & Management,* v. 16, no. 3, p. 291-305.

Webb, R.H., Ragland, H.C., Godwin, W.H., and Jenkins, D., 1978, Environmental effects of soil property changes with off road vehicle use: *Environmental Management,* v. 2, no. 3, p. 219–233.

Webb, R.H., and Wilshire, H.G., 1980, Recovery of soils and vegetation in a Mojave desert ghost town, Nevada, U.S.A.: *Journal of Arid Environments,* v. 3, p. 29 1–303.

Wemple, B.C., Jones, J., and Grant, G., 1996, *Channel network extension by logging roads in two basins, western Cascades,* Oregon: Water Resources Bulletin, v. 32, no. 6, p. 1195–1207.

Wheeler, G.L., and Rolfe, G.L., 1979, The relationship between daily traffic volume and the distribution of lead in roadside soil and vegetation: *Environmental Pollution,* v. 18, p. 265–274.

Wilcox, D.A., 1989, Migration and control of purple loosestrife (*Lythrum salicaria* L.) along highway corridors: *Environmental Management,* v. 13, no. 3, p. 365–370.

Wilkins, K.T., 1982, Highways as barriers to rodent dispersal: *Southwestern Naturalist,* v. 27, no. 44, p. 459–460.

Wilshire, H.G., 1983a, Off-road vehicle recreation management policy for public lands in the United States—A case history: *Environmental Management,* v. 7, no. 6, p. 489–499.

Wilshire, H.G., 1983b, The impact of vehicles on desert soil stabilizers, *in* Webb, R.H., and Wilshire, H.G., eds., *Environmental effects of off-road vehicles—Impacts and management in arid regions:* New York, Springer-Verlag, p. 31–50.

Winner, W.E., and Atkinson, C.J., 1986, Absorption of air pollutants by plants and consequences: *Trends in Ecology and Evolution,* v. 1, p. 15–18.

Yahner, R.H., 1988, Changes in wildlife communities near edges: *Conservation Biology,* v. 2, no. 4, p. 333–339.

Yahner, R.H., Morrell, T.E., and Rachael, J.S., 1989, Effects of edge contrast on depredation of artificial avian nests: *Journal of Wildlife Management,* v. 53, no. 4, p. 1135–1138.

Yanes, M., Velasco, J.M., and Suárez, F., 1995, Permeability of roads and railways to vertebrates—The importance of culverts: *Biological Conservation,* v. 71, no. 33, p. 2 17–222.

Appendix 1. Extensive Bibliographies

1.1. OHV Effects on Soils and Watersheds

Adams, J.A., Endo, A.S., Stolzy, L.H., Rowlands, P.G., and Johnson, H.B., 1982, Controlled experiments on soil compaction produced by off-road vehicles in the Mojave Desert, California: *Journal of Applied Ecology*, v. 19, no. 1, p. 167–175.

Adams, J.A., Stolzy, L.H., Endo, A.S., Rowlands, P.G., and Johnson, H.B., 1982, Desert soil compaction reduces annual plant cover: *California Agriculture,* v. 36, no. 9–10, p. 6–7.

Agrawal, Y.K., Patel, M.P., and Merh, S.S., 1981, Lead in soils and plants—Its relationship to traffic volume and proximity to highway (Lalbag, Baroda City): *International Journal of Environmental Studies,* v. 16, no. 3–4, p. 222–224.

Al-Awadhi, J.I., 2001, Impact of gravel quarrying on the desert environment of Kuwait: *Environmental Geology,* v. 41, no. 3–4, p. 365–371.

Albrecht, J., and Knopp, T.B., 1985, Off road vehicles—Environmental impact—Management response—A bibliography: Rosemount, Minnesota, *Agricultural Experiment Station,* University of Minnesota, Miscellaneous Publication, 50 p.

Amezketa Lizarraga, E., Gazol Lostao, R., and Aragues Lafarga, R., 2002, Development of an automated infiltrometer and its applicability to the field: *Investigacion Agraria Produccion y Proteccion Vegetales,* v. 17, no. 1, p. 13 1–142.

Anders, F.J., and Leatherman, S.P., 1987, Disturbance of beach sediment by off-road vehicles: *Environmental Geology and Water Sciences,* v. 9, no. 3, p. 183–189.

Anders, F.J., and Leatherman, S.P., 1987, Effects of off-road vehicles on coastal foredunes at Fire Island, New York, USA: *Environmental Management,* v. 11, no. 1, p. 45–52.

Arndt, W., 1996, The effect of traffic compaction on a number of soil properties: *Journal of Agriculture Engineering Research,* v. 11, p. 182–187.

Becher, H.H., 1985, Compaction of arable soils due to reclamation or off-road military traffic: *Reclamation and Revegetation Research,* v. 4, no. 2, p. 155–164.

Been, A., 1985, *Assessment of damage by wheels of off-road vehicles:* Norsk Institutt for Skogforskning, Technical Report, 12 p.

Beije, H.M., 1986, Effects of military-training activities on soil, vegetation, and fauna: Leersum, Netherlands, *Research Institute for Nature Management,* p. 95–111.

Bekker, M.G., 1977, *Evaluation of tires in off the road locomotion—Motion resistance and soil compaction:* Zesz Probl Postepow Nauk Roln, v. 183, p. 111–124.

Bekker, M.G., 1980, Evaluation of soil/vehicle relationship to lessen damage to forest road and off-road surfaces—A literature and state-of-the-art survey: San Dimas, Mexico, *Equipment Development Center,* Technical Report, 71 p.

Belnap, Jayne, 1993, Recovery rates of cryptobiotic crusts—Inoculant use and assessment methods: *Great Basin Naturalist,* v. 53, no. 1, p. 89–95.

Belnap, Jayne, 1995, Surface disturbances—Their role in accelerating desertification: *Environmental Monitoring and Assessment,* v. 37, no. 1–3, p. 39–57.

Belnap, Jayne, 2002, Impacts of off-road vehicles on nitrogen cycles in biological soil crusts— Resistance in different U.S. deserts: *Journal of Arid Environments,* v. 52, no. 2, p. 155–165.

Belnap, Jayne, 2003, The world at your feet—Desert biological soil crusts: Frontiers in *Ecology and the Environment,* v. 1, no. 4, p. 181–1 89.

Benoit, O., and Gotteland, P., 2005, Modelling of sinkage tests in tilled soils for mobility study: *Soil and Tillage Research,* v. 80, no. 1–2, p. 215–23 1.

Berry, K.H., 1980, A review of the effects of off-road vehicles on birds and other vertebrates, *in* DeGraaf, R.M., and Tilghman, N.G., eds., *Management of western forests and grasslands for nongame birds—Workshop proceedings,* Salt Lake City, Utah, February 11–14, 1980: Ogden Utah, U.S. Forest Service, Intermountain Forest and Range Experiment Station, General Technical Report INT–86, p. 451–467.

Bessee, G.B., and Kohl, K.B., 1993, *Characterization of CONUS and Saudi Arabian fine-grained soil samples:* San Antonio, Texas, Southwest Research Institute Belvoir Fuels and Lubricants Research Facility, Report no. BFLRF294, 80 p.

Boer, B., 1998, Anthropogenic factors and their potential impacts on the sustainable development of Abu Dhabi's terrestrial biological resources: *International Journal of Sustainable Development and World Ecology,* v. 5, no. 2, p. 125–135.

Bolling, J.D., and Walker, L.R., 2000, Plant and soil recovery along a series of abandoned desert roads: *Journal of Arid Environments,* v. 46, no. 11, p. 1–24.

Brainard, J., 1998, Patton Tank marks suggest long recovery: *Science News*, v. 154, no. 6, p. 87.

Braunack, M.V., 1986, The residual effects of tracked vehicles on soil surface properties: *Journal of Terramechanics,* v. 23, no. 1, p. 37–50.

Bredberg, C.J., and Waesterlund, I., 1983, Vehicle-caused damage to roots and soil: *Forstwissenschaftliches Centralblatt,* v. 102, no. 2, p. 86–98.

Brown, A.C., and McLachlan, A., 2002, Sandy shore ecosystems and the threats facing them— Some predictions for the year 2025: *Environmental Conservation,* v. 29, no. 1, p. 62–77.

Brown, G., and Porembski, S., 2000, Phytogenic hillocks and blow-outs as 'safe sites' for plants in an oil-contaminated area of northern Kuwait: *Environmental Conservation,* v. 27, no. 3, p. 242–249.

Brown, G., and Schoknecht, N., 2001, Off-road vehicles and vegetation patterning in a degraded desert ecosystem in Kuwait: *Journal of Arid Environments,* v. 49, no. 2, p. 413–427.

Brown, J., and Grave, N.A., 1979, *Physical and thermal disturbance and protection of permafrost:* Hanover, New Hampshire, Cold Regions Research and Engineering Lab, Report no. CRREL–SR–79–5, 46 p.

Bulinski, J., 2000, Effect of forward speed and kind of the agricultural aggregates on soil compaction by wheels: *Inzynieria Rolnicza,* v. 6, p. 111–117.

Burger, J., 1986, The effect of human activity on shorebirds in two coastal bays in northeastern United States: *Environmental Conservation,* v. 13, no. 2, p. 123–130.

Byrd, D.S., Gilmore, J.T., and Lea, R.H., 1983, Effect of decreased use of lead in gasoline on the soil of a highway: *Environmental Science and Technology,* v. 17, p. 12 1–123.

California State Water Resources Control Board, 1980, *Lake Tahoe Basin Water Quality Plan,* Final Plan: Sacramento, California, California State Water Resources Control Board, Technical Report.

Call, C.A., Barker, J.R., and McKell, C.M., 1981, Visitor impact assessment of scenic view areas at Bryce Canyon National Park: *Journal of Soil and Water Conservation,* v. 36, no. 1, p. 50–53.

Canfield, T.R., and Murray, M.J., 1992, *Use of finite elements to model soil/track interactions in coupled multi-body dynamic simulations—Is real time simulation feasible: Argonne National Laboratory,* Argonne, Illinois, Report no. ANL/CP–761 84, 3 p.

Charman, D.J., and Pollard, A.J., 1995, Long-term vegetation recovery after vehicle track abandonment on Dartmoor, SW England, U.K.: *Journal of Environmental Management,* v. 45, no. 1, p. 73–85.

Clampitt, C.A., 1993, Effects of human disturbances on prairies and the regional endemic *Aster curtus* in western Washington: *Northwest Science,* v. 67, no. 3, p. 163–169.

Cole, D.N., 1990, Trampling disturbance and recovery of cryptogamic soil crusts in Grand Canyon National Park: *Great Basin Naturalist,* v. 50, no. 4, p. 32 1–325.

Collins, E., O'Farrell, T.P., and Rhoads, W., 1982, *Annotated bibliography for biologic overview for the Nevada nuclear waste storage investigations,* Nevada Test Site, Nye County, Nevada: Goleta, California, EG and G, Inc., Report no. EGG1 1832419, 48 p.

Collins, E., O'Farrell, T.P., and Rhoads, W., 1982, *Biologic overview for the Nevada nuclear waste storage investigations,* Nevada Test Site, Nye County, Nevada: Goleta, California, EG and G, Inc., Report no. EGG1 1832460, 55 p.

Davidson, E.D., and Fox, M., 1974, *Effects of off-road motorcycle activity on Mojave Desert vegetation and soil:* Madroño, v. 22, no. 8, p. 381–390.

Davis, J.N., Baier, J., and McDonald, T., 2004, *Assessment of soil amendments for erosion control on off-road vehicle trails, 2004 annual international meeting of the Society for Engineering in Agricultural, Food and Biological Systems,* Ottawa, Canada, August 1–4, 2004: St. Joseph, Michigan, American Society of Agricultural Engineers.

Decker, D.J., Krueger, R.A., Bauer, R.A., Jr., Knuth, B.A., and Richmond, M.E., 1996, From clients to stakeholders—*A philosophical shift for fish and wildlife management:* Human Dimensions of Wildlife, v. 1, no. 1, p. 70-82.

Deletic, A., Ashley, R., and Rest, D., 2000, Modelling input of fine granular sediment into drainage systems via gully-pots: *Water Research,* v. 34, no. 15, p. 3836–3844.

Deliman, N.C., 1998, Modeling soil-traction element interaction for off-road mobility assessment with application to virtual environment simulation, American Society of Agricultural Engineers (ASAE) *International Meeting,* Orlando, Florida, March 8–10, 1998: St. Joseph, Michigan, American Society of Agricultural Engineers.

Della-Moretta, L.B., and Hodges, H.C., 1986, *Off-highway tire/road damage and healing mechanisms:* U.S. Department of Agriculture Forest Service Equipment Development Center, Report no. ASAE/TP86/1060, 15 p.

Duever, M.J., Riopelle, L.A., and McCollom, J.M., 1986, *Long term recovery of experimental off-road vehicle impacts and abandoned old trails in the Big Cypress National Preserve:* Naples, Florida, National Audubon Society, Technical Report, 56 p.

Dunn, J.P., Summerfield, C.J., and Johnson, M., 2003, Distribution, seasonal cycle, host plant records, and habitat evaluation of a Michigan threatened insect—The Great Plains spittlebug, *Lepyronia gibbosa* (Homoptera: Cercopidae): *Great Lakes Entomologist,* v. 35, no. 2, p. 121–129.

Dunnell, C.W., 1980, Protecting and rehabilitating ORV off-road recreational vehicles use areas—Erosion control and management, Wenatchee National Forest, in Andrews, R.N.L., and Nowak, P.F., eds., *Off-road vehicle use—A management challenge:* Washington, D.C., U.S. Department of Agriculture Office of Environmental Quality, p. 100–102.

Dyke, L.D., 1985, *Terrain disturbance due to summer off-road vehicle use in central Keewatin,* Northwest Territories, Canada: Ottawa, Ontario, Department of Indian Affairs and Northern Development, Report no. SSC–R71–19/36–1985F, 53 p.

Eckert, R.E.J., Wood, M.K., Blackburn, W.H., and Peterson, F.F., 1979, Impacts of off-road vehicles on infiltration and sediment production of two desert soils: *Journal of Range Management,* v. 32, no. 5, p. 394–397.

Erel, Y., 1998, Mechanisms and velocities of anthropogenic Pb migration in Mediterranean soils: *Environmental Research, v.* 78, no. 2, p. 112–117.

Falkengren-Grerup, U., 1986, Soil acidification and its impact on ground vegetation: *Oecologia,* v. 70, p. 339–347.

Fang, S., Wente, S., Gertner, G.Z., Wang, G., and Anderson, A., 2002, Uncertainty analysis of predicted disturbance from off-road vehicular traffic in complex landscapes at Fort Hood: *Environmental Management,* v. 30, no. 2, p. 199–208.

Fatoki, O.S., 2000, Trace zinc and copper concentrations in roadside vegetation and surface soils—A measurement of local atmospheric pollution in Alice, South Africa: *International Journal of Environmental Studies* (UK), v. 57, no. 5, p. 501–513.

Fletcher, J.J., and Lovejoy, S.B., 1986, Off-farm costs of sediment from agricultural lands, *Proceedings of the American Society of Agricultural Engineers,* Winter Meeting, Chicago, Illinois: St. Joseph, Michigan, American Society of Agricultural Engineers.

Flint, A.L., and Childs, S., 1984, Development and calibration of an irregular hole bulk density sampler: *Soil Science Society of America Journal,* v. 48, no. 2, p. 374–378.

Folz, R.B., 2006, Erosion from all terrain vehicle (ATV) trails on National Forest lands (abs.): St. Joseph, Michigan, Proceedings of the 2006 American Society of Agricultural and Biological Engineers, *http://asae.frymulti.com/abstract.asp?aid=21056&t=2.*

Foresman, C.L., 1976, Effect of snowmobile traffic on bluegrass (*Pao pratensis*): *Journal of Environmental Quality,* v. 5, no. 2, p. 129–131.

Freitag, D.R., Green, A.J., and Murphy, N.R., 1964, *Normal stresses at the tire-soil interface in yielding soils:* Vicksburg, Mississippi, U.S. Army Engineer Waterways Experiment Station, Report no. AEWES–MISC–PAPER–4–629, 34 p.

Fuchs, E.H., Wood, M.K., Jones, T.L., and Racher, B., 2003, Impacts of tracked vehicles on sediment from a desert soil: *Journal of Range Management,* v. 56, no. 4, p. 342–352.

Furniss, M.J., Flanagan, S.A., and McFadin, B.A., 2000, *Hydrologically connected roads—An indicator of the influence of roads on chronic sedimentation, surface water hydrology, and exposure to toxic chemicals:* U.S. Forest Service, Stream Systems Technology

Center, Rocky Mountain Research Station, Technical Report, 4 p., *http://www.stream.fs.fed.us/streamnt/jul100/jul100_2.htm.*

Gardner, R.B., 1979, Some environmental and economic effects of alternative forest road designs: *Transactions of the American Society of Agricultural Engineers,* v. 22, no. 1, p. 63–68.

Geological Society of America, 1977, *Impacts and management of off-road vehicles:* Boulder, Colorado, Report of the Committee on Environment and Public Policy, Technical Report, 8 p.

Gilbertson, D., 1983, The impacts of off-road vehicles in the Coorong Dune and Lake Complex of South Australia, in Webb, R.H., and Wilshire, H.G., eds., *Environmental effects of off- road vehicles—Impacts and management in arid regions:* New York, Springer-Verlag, p. 355–373.

Gillette, D.A., and Adams, J., 1983, Accelerated wind erosion and prediction rates of off-road vehicles, in Webb, R.H., and Wilshire, H.G., eds., *Environmental effects of off-road vehicles—Impacts and management in arid regions:* New York, Springer-Verlag, p. 97–109.

Gish, C.D., and Christensen, R.E., 1973, *Cadmium, nickel, lead, and zinc in earthworms from roadside soil: Environmental Science and Technology,* v. 7, p. 1060–1062.

Gray, J.R., 1977, *Kinds and costs of recreational pollution in the Sandia Mountains: New Mexico Agricultural Experiment Station Bulletin,* Technical Report, 57 p.

Green, J.E., and Knight, S.J., 1959, *Preliminary study of stresses under off-road vehicles:* Vicksburg, Mississippi, U.S. Army Engineer Waterways Experiment Station, Report no. AEWES–MISC–PAPER–4–362, 20 p.

Griggs, G.B., and Walsh, B.L., 1981, The impact, control, and mitigation of off-road vehicle activity in Hungry Valley, California: *Environmental Geology,* v. 3, no. 4, p. 229–243.

Grimes, D.W., Miller, R.J., and Wiley, P.L., 1975, Cotton and corn root development in two field soils of different strength characteristics: *Agronomy Journal,* v. 67, p. 5 19–523.

Grimes, D.W., Sheesley, W.R., and Wiley, P.L., 1978, Alfalfa root development and shoot regrowth in compact soil of wheel traffic patterns: *Agronomy Journal,* v. 70, p. 955–958.

Hairsine, P.B., Croke, J.C., Mathews, H., Fogarty, P., and Mockler, S.P., 2002, Modelling plumes of overland flow from logging tracks: *Hydrological Processes*, v. 16, no. 12, p. 23 11–2327.

Hall, C., and Dearden, P., eds., 1984, *The impact of "non-consumptive" recreation on wildlife— An annotated bibliography*: Monticello, Illinois, Vance Bibliographies, 45 p.

Hammitt, W.E., and Cole, D.N., 1987, *Wildland recreation—Ecology and management:* New York, John Wiley & Sons, 341 p.

Hansen, D.J., and Ostler, W.K., 2001, *Plant-damage assessment technique for evaluating military vehicular impacts to vegetation in the Mojave Desert: Las Vegas,* Nevada, Bechtel Nevada, Inc., Report no. DOE/NV /1178613.

Harrison, R.T., 1980, Environmental impact of off-road motorcycles, in Andrews, R.N.L., and Nowak, P.F., eds., *Off-road vehicle use—A management challenge:* Washington, D.C., U.S. Department of Agriculture Office of Environmental Quality, p. 266–269.

Haupt, H.F., 1959, *Road and slope characteristics affecting sediment movement from logging roads: Journal of Forestry*, v. 57, no. 5, p. 329–339.

Heede, B.H., 1983, Control of rills and gullies in off-road vehicle traffic areas, in Webb, R.H., and Wilshire, H.G., eds., *Environmental effects of off-road vehicles—Impacts and management in arid regions:* New York, Springer-Verlag, p. 245–264.

Hellstroem, G.B., 1996, Preliminary investigations into recent changes of the Goukamma Nature Reserve frontal dune system, South Africa—With management implications: *Landscape and Urban Planning,* v. 34, no. 3–4, p. 225–23 5.

Hinckley, B.S., Iverson, R.M., and Hallet, B., 1983, Accelerated water erosion in ORV off-road vehicle-use areas, in Webb, R.H., and Wilshire, H.G., eds., *Environmental effects of off-road vehicles—Impacts and management in arid regions:* New York, Springer-Verlag, p. 8 1–96.

Hirst, R.A., Pywell, R.F., Marrs, R.H., and Putwains, P.D., 2003, The resistance of a chalk grassland to disturbance: *Journal of Applied Ecology,* v. 40, no. 2, p. 3 68–379.

Holloway, D.C., Wilson, W.H., and Drach, T.J., 1989, *Examination of ATV tire forces generated on clay, grass, and sand surfaces:* Warrendale, Pennsylvania, Society of Automotive Engineers, Technical Report, 12 p.

Holsman, R.H., 2005, Management opportunities and obligations for mitigating off-road vehicle impacts to wildlife and their habitats: *Transactions of the North American Wildlife and Natural Resources Conference,* v. 70, p. 399–417.

Hooks, C.L., and Jansen, I.J., 1986, Recording cone penetrometer developed in reclamation research: *Soil Science Society of America Journal,* v. 50, no. 1, p. 10–12.

Hosier, P.E., and Eaton, T.E., 1980, *The impact of vehicles on dune and grassland vegetation on a southeastern North Carolina barrier beach:* Wilmington, North Carolina, North Carolina University at Wilmington, Report no. UNC–SG–R–1 64, 13 p.

Huszar, P.C., and Piper, S.L., 1986, Estimating the off-site costs of wind erosion in New Mexico: *Journal of Soil and Water Conservation,* v. 41, no. 6, p. 414–421.

Hyatt, J.A., and Gilbert, R., 2000, Lacustrine sedimentary record of human-induced gully erosion and land-use change at Providence Canyon, southwest Georgia, USA: *Journal of Paleolimnology,* v. 23, no. 4, p. 421–438.

Hyers, A.D., and Marcus, M.G., 1981, Land use and desert dust hazards in central Arizona, in Pewe, T., ed., *Desert dust origin, characteristics, and effect on man: Boulder, CO, Geological Society of America,* Inc., p. 267–280.

International Society for Terrain-Vehicle Systems, 1984, *Proceedings of the international conference on the performance of off-road vehicles and machines,* Cambridge, England, August 5–11, 1984: International Society for Terrain-Vehicle Systems, 384 p.

International Society for Terrain-Vehicle Systems, 1994, *Proceedings of the European ISTVS conference (6th), OVK symposium (4th), on off road vehicles in theory and practice, Vienna, Austria, September 28–30, 1994:* International Society for Terrain-Vehicle Systems, 349 p.

Isensee, E., and Luth, H.G., 1992, *Continuous recording of soil density:* Landtechnik, v. 47, no. 9, p. 449–452.

Iverson, R.M., 1980, Processes of accelerated pluvial erosion on desert hillslopes modified by vehicular traffic: *Earth Surface Processes,* v. 5, p. 369-3 88.

Iverson, R.M., Hinckley, B.S., and Webb, R.M., 1981, Physical effects of vehicular disturbances on arid landscapes: *Science,* v. 212, no. 4497, p. 915–917.

Johnson, C.W., and Smith, J.P., 1983, Soil loss caused by off-road vehicle use on steep slopes: *Transactions of the American Society of Agricultural Engineers,* v. 26, no. 2, p. 402–405.

Johnson, R.R., Mills, G.S., and Carothers, S.W., 1990, Creation and restoration of riparian habitat in southwestern arid and semi-arid regions, in Kusler, J.A., and Kentula, M.E., eds., Wetland creation and restoration—*The status of the science:* Covelo, California, Island Press, p. 351–366.

Jones, R., Horner, D., Sullivan, P., and Ahlvin, R., 2005, A methodology for quantitatively assessing vehicular rutting on terrains: *Journal of Terramechanics,* v. 42, no. 3–4, p. 245–257.

Karafiath, L.L., 1978, *Track-soil interaction model for the determination of maximum soil thrust*: Bethpage, New York, Grumman Aerospace Corporation, Report no. RE–556, 60 p.

Kay, B.L., and Graves, W.L., 1983, History of revegetation studies in the California deserts, in Webb, R.H., and Wilshire, H.G., eds., *Environmental effects of off-road vehicles—Impacts and management in arid regions*: New York, Springer-Verlag, p. 315–324.

Kay, B.L., and Graves, W.L., 1983, Revegetation and stabilization techniques for disturbed desert vegetation, in Webb, R.H., and Wilshire, H.G., eds., *Environmental effects of off-road vehicles—Impacts and management in arid regions:* New York, Springer-Verlag, p. 3 25–340.

Kay, J., 1981, Evaluating environmental impacts of off-road vehicles: *Journal of Geography,* v. 80, no. 1, p. 10–18.

Keener, K.M., Wood, R.K., Holmes, R.G., and Morgan, M.T., 1991, *Soil strength evaluation of sample cores in a field measurement system:* : St. Joseph, Michigan, American Society of Agricultural Engineers, ASAE Paper No. 91-1526, p. 10.

Kennedy, J.G., Collins, J.G., and Smith, M.H., 1967, *Moisture—Strength characteristics of selected soils in Thailand. Volume I—Analyses and application of data: Vicksburg, Mississippi, U.S.* Army Engineer Waterways Experiment Station, Report no. AEWES–TR–3–79 1–VOL–1, 126 p.

Kert, J., 1992, *Remote sensing of off-road vehicle emissions, International conference optical remote sensing and applications to environmental and industrial safety problems,* Houston, Texas, April 6–8, 1992.

Khalaf, F.I., 1989, Desertification and aeolian processes in the Kuwait Desert: *Journal of Arid Environments,* v. 16, no. 2, p. 125–145.

Knott, J.M., 1978, *Reconnaissance assessment of erosion and sedimentation in the Canada de los Alamos Basin, Los Angeles and Ventura Counties,* California: U.S. Geological Survey, Open File Report no. 78–873, 49 p.

Komatsu, M., Kato, A., and Sakamoto, K., 1988, *Study on the automatic measurement system of soil parameters.* Development of soil hardness sensor: Tottori, Japan, Tottori University, Bulletin of the Faculty of Agriculture, Tottori University, v. 41, p. 39–46.

Kondolf, G.M., Piegay, H., and Landon, N., 2002, *Channel response to increased and decreased bedload supply from land use change—Contrasts between two catchments: Geomorphology,* v. 45, no. 1–2, p. 35–51.

Koppel, W., 1988, *Dynamic impact on soil structure due to traffic of off-road vehicles, in* Drescher, J., Horn, R., and Boodt, M.D., eds., Impact of water and external forces on soil structure—Selected papers of the 1st Workshop on Soilphysics and Soilmechanics: Cremlingen-Destedt, Germany, Catena Verlag, p. 113–122.

Kurczerski, F.E., 2000, *History of white pine (Pinus strobus)/oak (Quercus spp.) savanna in southern Ontario, with particular reference to the biogeography and status of the antenna- waving wasp, Tachysphex pechumani* (Hymenoptera: Sphecidae): Canadian Field-Naturalist, v. 114, no. 1, p. 1–20.

Kutiel, P., Eden, E., and Zhevelev, Y., 2000, Effect of experimental trampling and off-road motorcycle traffic on soil and vegetation of stabilized coastal dunes, Israel: *Environmental Conservation,* v. 27, no. 1, p. 14–23.

Kutiel, P., Eden, Z., and Zhevelev, H., 2001, The impact of motorcycle traffic on soil and vegetation of stabilized coastal dunes, Israel: *Journal of Coastal Conservation,* v. 7, no. 1, p. 81–89.

Lacey, R.M., and Severinghaus, W.D., 1982, *Evaluation of lands for off-road recreational four- wheel drive vehicle use:* U.S. Department of Commerce Government Reports Announcements and Index National Technical Information Service (NTIS), Technical Report, 1755 p.

Langdon, A.M., 2000, *Mojave Desert soils, plants, and ants—Developing a monitoring strategy for off-highway-vehicles:* Pomona, California, California State Polytechnic University, Pomona, 114 p.

Larney, F.J., Leys, J.F., Mueller, J.F., and McTainsh, G.H., 1999, Dust and endosulfan deposition in a cotton-growing area of northern New South Wales, Australia: *Journal of Environmental Quality,* v. 28, no. 2, p. 692–701.

Lawson, D.E., 1982, *Long-term modifications of perennially frozen sediment and terrain at East Oumalik, northern Alaska:* Hanover, New Hampshire, Cold Regions Research and Engineering Lab, Technical Report, 44 p.

Lei, S.A., 2004, *Soil compaction from human trampling, biking, and off-road motor vehicle activity in a blackbrush* (*Coleogyne ramosissima*) shrubland: Western North American Naturalist, v. 64, no. 1, p. 125–130.

Leis, S.A., Engle, D.M., Leslie, D.M., and Fehmi, J.S., 2005, Effects of short- and long-term disturbance resulting from military maneuvers on vegetation and soils in a mixed prairie area: *Environmental Management,* v. 36, no. 6, p. 849–861.

Lightfoot, D.C., and Whitford, W.G., 1991, *Productivity of creosotebush foliage and associated canopy arthropods along a desert roadside:* American Midland Naturalist, v. 125, no. 2, p. 310–322.

Liston, N., Hutt, M., and White, L., 1981, *Mobility bibliography: Hanover, New Hampshire, Cold Regions Research and Engineering Lab,* Report no. CRREL–SR–81–29, 333 p.

Liston, R.A., Czako, T., Haley, P., Harrison, W.L., and Hanamoto, B., 1966, *Mobility environmental research study mobility testing procedures:* Warren, Michigan, U.S. Army Tank-Automotive Center—Warren Michigan Land Locomotion Lab, Technical Report, 88 p.

Lodico, N.J., 1973, *Environmental effects of off-road vehicles—A review of the literature:* U.S. Office of Library Services, Research Services Branch, Technical Report, 109 p.

Lovich, J.E., and Bainbridge, D., 1999, Anthropogenic degradation of the southern California desert ecosystem and prospects for natural recovery and restoration: *Environmental Management*, v. 24, no. 3, p. 309–326.

Luckenbach, R.A., 1978, *An analysis of off- road vehicle use on desert avifaunas, Transactions of the North American Wildlife and Natural Resources Conference,* v. 43, p. 157–162.

Luckenbach, R.A., 1982, *Ecology and management of the desert tortoise (Gopherus agassizii) in California:* U.S. Fish and Wildlife Service, Technical Report, 37 p.

Luckenbach, R.A., and Bury, R.B., 1983, Effects of off-road vehicles on the biota of the Algodones Dunes, Imperial County, California, USA: *Journal of Applied Ecology,* v. 20, no. 1, p. 265–286.

Manson, D.A., 1980, Erosion control in Angourie National Park: *Journal of the Soil Conservation Service of New South Wales,* v. 36, no. 1, p. 23–32.

McBrayer, M.C., Mauldon, M., Drumm, E.C., and Wilson, G.V., 1997, Infiltration tests on fractured compacted clay: *Journal of Geotechnical and Geoenvironmental Engineering,* v. 123, no. 5, p. 469–473.

McCool, D.K., Dossett, M.G., and Yecha, S.J., 1981, Portable rill meter for field measurement of soil loss, Erosion and sediment transport measurement—Proceedings of the Florence symposium, Florence, Italy, 22-26 June 1981: Washington, D.C., *International Association of Hydrological Sciences,* p. 479–484.

Melvin, S.M., Griffin, C.R., and MacIvor, L.H., 1991, Recovery strategies for piping plovers in managed coastal landscapes: *Coastal Management,* v. 19, no. 1, p. 2 1–34.

Meyer, K.G., 2002, *Managing degraded off-highway vehicle trails in wet, unstable, and sensitive environments: U.S. National Park Service,* Report no. PB2005 105502, 88 p.

Miller, P.M., 1968, *The application of the visioplasticity method to soft-soil mobility problems:* Buffalo, New York, Cornell Aeronautical Laboratory and United States Advanced Research Projects Agency, p. 41.

Miller, R.E., Hazard, J., and Howes, S., 2001, *Precision, accuracy, and efficiency of four tools for measuring soil bulk density or strength:* U.S. Forest Service, Pacific Northwest Research Station, Report no. PNW–RP–532, 16 p.

Misak, R.F., Al-Awadhi, J.M., Omar, S.A., and Shahid, S.A., 2002, *Soil degradation in Kabd area, southwestern Kuwait City: Land Degradation and Development,* v. 13, no. 5, p. 403–415.

Mize, R., Evans, R.E., MacRoberts, B.R., MacRoberts, M.H., and Rudolph, D.C., 2005, Restoration of pitcher plant bogs in eastern Texas, USA: *Natural Areas Journal,* v. 25, no. 2, p. 197–201.

Moorhead, D.L., Linkins, A.E., and Everett, K.R., 1996, Road dust alters extracellular enzyme activities in tussock tundra soils, Alaska, U.S.A.: *Arctic and Alpine Research,* v. 28, no. 3, p. 346-351.

Motto, H.L., Daines, R.H., Chilko, D.M., and Motto, C.K., 1970, Lead in soils and plants—Its relationship to traffic volume and proximity to highways: *Environmental Science and Technology,* v. 4, p. 231–237.

Nakata, J.K., 1983, Off-road vehicular destabilization of hill slopes—The major contributing factor to destructive debris flows in Ogden, Utah, 1979, in Webb, R.H., and Wilshire, H.G., eds., *Environmental effects of off-road vehicles—Impacts and management in arid regions:* New York, Springer-Verlag, p. 343–353.

National Oceanic and Atmospheric Administration, 2005, Understanding the "human dimension" of coastal management using social science, *http://maps.csc.noaa.gov/socialscience_2/wheel_stakeholders.htm* (accessed April 2007).

Niedoroda, A., 1978, *The geomorphologic effects of off-road vehicles on coastal systems of Cape Cod, Massachusetts:* Amherst, Massachusetts, Massachusetts University National Park Service Cooperative Research Unit, Report no. UM–NPSCRU–17, 106 p.

Noor, M., and Shah, B.H., 1995, Infiltration capacity and soil bulk density differences between grazed area and off-road track at Paya (Kaghan Valley): *Pakistan Journal of Forestry,* v. 45, no. 1, p. 13–18.

Noren, O., Danfors, B., and Stambeck, A., 1984, *Technical systems in energy forestry:* Stockholm, Sweden, Statens Energiverk, Report no. STEV–EO–84–4, 145 p.

Noren, O., Danfors, B., Stambeck, A., and Gutekunst, K., 1982, *Technical equipment for the cultivation of energy forest:* Uppsala, Sweden, Jordbrukstekniska Institute, Report no. JTI–38, 116 p.

Nyssen, J., Poesen, J., Moeyersons, J., Luyten, E., Veyret-Picot, M., Deckers, J., Haile, M., and Govers, G., 2002, Impact of road building on gully erosion risk—*A case study from the northern Ethiopian Highlands: Earth Surface Processes and Landforms,* v. 27, no. 12, p. 1267–1283.

Okello, J.A., 1991, A review of soil strength measurement techniques for prediction of terrain vehicle performance: *Journal of Agricultural Engineering Research,* v. 50, no. 2, p. 129–155.

O'Sullivan, M.F., and Ball, B.C., 1982, A comparison of five instruments for measuring soil strength in cultivated and uncultivated cereal seedbeds: *Journal of Soil Science,* v. 33, no. 4, p. 597–608.

Payne, G.F., Leininger, W.D., and Foster, J., 1979, How off-road vehicles affect range quality. *When driving across rangelands, it's better to blaze a new trai*l: Bozeman, Montana, Montana Agricultural Experiment Station, Report, 2 p.

Perez, F.L., 1991, *Particle sorting due to off-road vehicle traffic in a high Andean paramo:* Catena Giessen, v. 18, no. 3–4, p. 239–254.

Persico, L.P., Nichols, K.K., and Bierman, P.R., 2005, *Tracking painted pebbles—Short-term rates of sediment movement on four Mojave Desert piedmont surfaces:* Water Resources Research, v. 41, no. 7, p. W07004.07001–W07004.07015.

Piehl, B.T., Beschta, R.L., and Pyles, M.R., 1988, Ditch-relief culverts and low-volume forest roads in the Oregon Coast Range: *Northwest Science,* v. 62, no. 3, p. 9 1–98.

Pinard, M.A., Barker, M.G., and Tay, J., 2000, Soil disturbance and post-logging forest recovery on bulldozer paths in Sabah, Malaysia: *Forest Ecology and Management,* v. 130, no. 1–3, p. 213–225.

Piper, S., and Huszar, P.C., 1989, Re-examination of the off-site costs of wind erosion in New Mexico: *Journal of Soil and Water Conservation,* v. 44, no. 4, p. 332–334.

Powers, R.F., Tiarks, A.E., and Boyle, J.R., 1998, Assessing soil quality—Practicable standards for sustainable forest productivity in the United States, *in* The contribution of soil science to the development of and implementation of criteria and indicators of sustainable forest management: St. Louis, Missouri, *Soil Science Society of America,* Inc., p. 53–80.

Prose, D.V., 1985, Persisting effects of armored military maneuvers on some soils of the Mojave Desert: *Environmental Geology and Water Sciences*, v. 7, no. 3, p. 163–170.

Prose, D.V., Metzger, S.K., and Wilshire, H.G., 1987, Effects of substrate disturbance on secondary plant succession—Mojave Desert, California: *Journal of Applied Ecology,* v. 24, no. 1, p. 305–313.

Qiu, X.D., Ji, X.W., and Zhuang, J.D., 1995, Analysis of desert sand properties related to off-road locomotion: *Transactions of the Chinese Society of Agricultural Engineering,* v. 11, no. 2, p. 11–16.

Quarles, H.D., Hanawalt, R.B., and Odum, W.E., 1974, Lead in small mammals, plants, and soil at varying distances from a highway: *Journal of Applied Ecology,* v. 11, no. 3, p. 937–949.

Rafique, S.A., 1994, Effect of grazing management and fertilizer application on vegetation and soil properties of a moist temperate forest range in Siran Valley (Mansehra), NWFP: *Pakistan Journal of Forestry,* v. 44, no. 1, p. 20–29.

Raghavan, G.S.V., 1977, Effect of wheel slip on soil compaction: *Journal of Agricultural Engineering Research,* v. 22, no. 1, p. 79–83.

Raghavan, G.S.V., McKyes, E., Amir, I., Chasse, M., and Broughton, R.S., 1976, Prediction of soil compaction due to off-road vehicle traffic: *Transactions of the American Society of Agricultural Engineers,* v. 19, no. 4, p. 610–613.

Richard, G., Tan, R., and Avery, J., 1977, *An implementation plan for suspended particulate matter in the Phoenix area.* Volume II. Emission inventory: U.S. Environmental Protection Agency, Office of Air Quality Planning and Standards, Report no. EPA450377021B, 218 p.

Rickard, W.H., 1988, *Natural vegetation at the proposed reference repository location in southeastern Washington:* Richland, Washington, Battelle Pacific Northwest Labs, Report no. PNL–6402, 27 p.

Ross, J.B., and Willison, J.H.M., 1991, Impacts of all-terrain vehicles on bogs and barrens of the Cape Breton Highlands—*Proceedings of the International Conference on Science and Management of Protected Areas,* Nova Scotia, Canada, 14-19 May 1991: Halifax, Nova Scotia, Canada, p. 533–534.

Rula, A.A., Freitag, D.R., and Knight, S.J., 1966, Concepts for vehicles for off-road use in remote areas: Vicksburg, Mississippi, *U.S. Army Engineer Waterways Experiment Station,* Report no. AEWESMISCPAPER4854.

Rula, A.A., and Nuttall, C.J., 1971, An analysis of ground mobility models (ANAMOB): Vicksburg, Mississippi, U.S. Vicksburg, Mississippi, *U.S. Army Engineer Waterways Experiment Station,* Report no. AEWESTRM714, 326 p.

Samaras, Z., and Zierock, K.H., 1995, Off-road vehicles—*A comparison of emissions with those from road transport: Science of the Total Environment,* v. 169, p. 249–253.

Schemnitz, S.D., and Schortemeyer, J.D., 1972, *The influence of vehicles on Florida Everglades vegetation:* Tallahassee, Florida, Florida State Game and Fresh Water Fish Commission, Report no. DISFEP–74–3 1, 65 p.

Schreier, H., and Lavkulich, L.M., 1979, *A numerical approach to terrain analysis for off-road trafficability: Photogrammetric Engineering and Remote Sensing,* v. 45, no. 5, p. 635–642.

Schultink, G., 1977, *Impact analysis of off-road-vehicle use on vegetation in the Grand Mere Dune environment:* East Lansing, Michigan, Michigan State University, Report no. NASACR155764, 10 p.

Severinghaus, W.D., Riggins, R.E., and Goran, W.D., 1980, *Effects of tracked vehicle activity on terrestrial mammals and birds at Fort Knox, Kentucky:* Transactions of the Kentucky Academy of Science, v. 41, no. 1–2, p. 15–26.

Shaaban, S., 1984, *Compaction of sand using ordinary off-road vehicles:* Cairo, Egypt, Military Technical College, Report no. ADP0042978, 11 p.

Shay, R., 1979, *Management problems in off-road-vehicle recreation,* in Ittner, R., Potter, D.R., Agee, J.K., and Anschell, S., eds., Recreational impact on wildlands—Conference proceedings, Seattle, Washington, October 27–29, 1978: U.S. Forest Service, Pacific Northwest Forest and Range Experiment Station and U.S. National Park Service, p. 314–317.

Shikanai, T., Ueno, M., Hashiguchi, K., Nohse, Y., and Okayasu, T., 1997, A precise measurement of soil deformation under the wheel: *Journal of the Japanese Society of Agricultural Machinery,* v. 59, no. 2, p. 3–11.

Shoop, S.A., 1993, *Terrain characterization for trafficability*: Hanover, New Hampshire, Cold Regions Research and Engineering Lab, Report no. CRREL–93–6, 29 p.

Snyder, C.T., Frickel, D.G., Hadley, R.F., and Miller, R.F., 1976, *Effects of off-road vehicle use on the hydrology and landscape of arid environments in central and southern California:* U.S. Geological Survey Water Resources Investigations Report no. USGS/WRI–76–99, 52 p.

Sparrow, S.D., Wooding, F.J., and Whiting, E.H., 1978, Effects of off-road vehicle traffic on soils and vegetation in the Denali Highway region of Alaska: *Journal of Soil and Water Conservation,* v. 33, no. 1, p. 20–27.

Spellerburg, I.F., and Morrison, T., 1998, *The ecological effects of new roads—A literature review:* Wellington, New Zealand, New Zealand Department of Conservation, Technical Report, 55 p.

Startsev, A.D., and Mcnabb, D.H., 2000, Effects of skidding on forest soil infiltration in west-central Alberta: *Canadian Journal of Soil Science,* v. 80, no. 4, p. 6 17–624.

Steiner, A.J., and Leatherman, S.P., 1979, *An annotated bibliography of the effects of off-road vehicle and pedestrian traffic on coastal ecosystems: U.S. National Park Service Cooperative Research Unit, and University of Massachusetts Amherst,* Report no. UMNPSCRU45, 88 p.

Steiner, A.J., and Leatherman, S.P., 1981, Recreational impacts on the distribution of ghost crabs *Ocypode quadrata* Fab.: *Biological Conservation,* v. 20, no. 2, p. 111–122.

Stephenson, G., 1999, Vehicle impacts on the biota of sandy beaches and coastal dunes—A review from a New Zealand perspective: *Science for Conservation,* v. 121, p. 1–14.

Stinson, B., 1971, *An analytical model for predicting cross-country vehicle performance. Appendix E-quantification of the screening effects of vegetation on driver's vision and vehicle speed:* Vicksburg, Mississippi, U.S. Army Engineer Waterways Experiment Station, Report no. AEWESTR3783, 33 p.

Stroh, T., 2001, *Proposed Collbran rock and soil source project—Wildlife report:* U.S. Bureau of Reclamation, Report no. PB2006101846, 28 p.

Stull, R., Shipley, Susan, Hovanitz, E., Thompson, S., and Hovanitz, K., 1979, *Effects of off-road vehicles in Ballinger Canyon,* California: Geology, v. 7, no. 1, p. 19–21.

Swanson, G.D., 1971, *Studies of dual and tandem rigid wheel performance in sand:* Hoboken, New Jersey, Stevens Institute of Technology, Report no. SIT–DL–71–1536, 132 p.

Tunstall, B.R., and Reece, P.H., 1989, *Environmental assessment of the Sunset and Big Desert lands,* northwest Victoria, Australia: South, Australia, Australia Commonwealth

Scientific and Industrial Research Organization, Division of Water Resources Divisional Report, Clayton Technical Report, 1–85 p.

Turnage, G.W., 1972, Performance of soils under tire loads. Report 8. *Application of test results to tire selection for off-road vehicles:* Vicksburg, Mississippi, U.S., Army Engineer Waterways Experiment Station, Report no. AEWES–TR–3–666–8, 164 p.

Turton, S.M., 2005, Managing environmental impacts of recreation and tourism in rainforests of the wet tropics of Queensland World Heritage Area: *Geographical Research,* v. 43, no. 2, p. 140–151.

Tuttle, M., and Griggs, G., 1985, Accelerated soil erosion at three State Vehicular Recreation Areas, central and southern California, Erosion control—A challenge in our time—*Proceedings of the 16th annual International Erosion Control Association, San Francisco, California,* February 2 1–22, 1985: Steamboat Springs, Colorado, p. 105-115.

Tuttle, M., and Griggs, G., 1987, Soil erosion and management recommendations at three State Vehicular Recreation Areas, California: *Environmental Geology and Water Sciences,* v. 10, no. 2, p. 111–123.

Twiss, R., Sidener, J., Bingham, G., Burke, J.E., and Hall, C.H., 1980, *Potential impacts of geothermal development on outdoor recreational use of the Salton Sea:* Livermore, California, University of California, Livermore, Technical Report, 61 p.

U.S. Army Corps of Engineers, 1994, *Stationing of mechanized or armored combat forces at Fort Lewis, Thurston and Pierce Counties,* Washington: U.S. Army Corps of Engineers, Report no. 940045.

U.S. Army Corps of Engineers, 2001, *Fort Bliss mission and master plan—Dona Ana and Otero Counties,* New Mexico, and El Paso County, Texas: U.S. Army Corps of Engineers, Report no. 010081.

U.S. Army Test and Evaluation Command, 1970, *Soft-soil vehicle mobility: Aberdeen Proving Ground,* Maryland, U.S. Army Report no. MTP–2–2–619, 11 p.

U.S. Bureau of Land Management, 1984, *Proposed Monument resource management plan, Idaho:* U.S. Bureau of Land Management, Report no. 840584, 419 p.

U.S. Bureau of Land Management, 1985, *Box Elder resource management plan,* Utah: U.S. Bureau of Land Management, Report no. 850448.

U.S. Bureau of Land Management, 1985, *Grand Junction resource area—Resource management plan and environmental impact statement:* U.S. Bureau of Land Management, Technical Report, 171 p.

U.S. Bureau of Land Management, 1985, *resource management plan for the Walker Planning Area, Nevada:* U.S. Bureau of Land Management, Report no. 850274.

U.S. Bureau of Land Management, 1985, *White Sands resource area management plan, New Mexico:* U.S. Bureau of Land Management, Report no. 850387.

U.S. Bureau of Land Management, 1986, *Lander resource management plan,* Lander, Wyoming: U.S. Bureau of Land Management, Report no. 860449.

U.S. Bureau of Land Management, 1986, *Little Snake resource management plan,* Moffat, Rio Blanco, and Routt Counties, Colorado: U.S. Bureau of Land Management, Report no. 860029.

U.S. Bureau of Land Management, 1986, *Proposed 1985 amendments to the California desert conservation area plan and the Eastern San Diego County master framework plan,* California: U.S. Bureau of Land Management, Report no. 860432, 138 p.

U.S. Bureau of Land Management, 1986, *Proposed 1985 amendments to the California desert conservation area plan,* California: U.S. Bureau of Land Management, Report no. 860077.

U.S. Bureau of Land Management, 1986, *resource management plan for the House Range resource area, Juab and Millard Counties,* Utah: U.S. Bureau of Land Management, Report no. 860377.

U.S. Bureau of Land Management, 1986, *resource management plan for the Warm Springs resource area,* Millard County, Utah: U.S. Bureau of Land Management, Report no. 860395.

U.S. Bureau of Land Management, 1986, *Wilderness Suitability EIS—Grass Creek and Cody resource areas, Bighorn Basin, Wyoming* (draft supplement for the Owl Creek Wilderness Study Area): U.S. Bureau of Land Management, Report no. 860168, 40 p.

U.S. Bureau of Land Management, 1987, *Cascade resource management plan, Idaho: U.S. Bureau of Land Management,* Report no. 870281, 432 p.

U.S. Bureau of Land Management, 1987, *Medicine Bow-Divide resource areas resource management plan,* Rawlins District, Wyoming: U.S. Bureau of Land Management, Report no. 870212.

U.S. Bureau of Land Management, 1987, *North Dakota resource management plan, Dunn and Bowman Counties,* North Dakota: U.S. Bureau of Land Management, Report no. 870236, 199 p.

U.S. Bureau of Land Management, 1987, *Pinedale resource area resource management plan, Sublette and Lincoln Counties*, Wyoming: U.S. Bureau of Land Management, Report no. 870077, 347 p.

U.S. Bureau of Land Management, 1987, *Pinedale resource area resource management plan,* Sublette, Teton, Lincoln, and Fremont Counties, Wyoming: U.S. Bureau of Land Management, Report no. 870428, 232 p.

U.S. Bureau of Land Management, 1987, *Proposed resource management plan for the San Juan resource area,* Moab District, Utah: U.S. Bureau of Land Management, Report no. 870440.

U.S. Bureau of Land Management, 1987, *Taos resource management plan, New Mexico:* U.S. Bureau of Land Management, Report no. 870331.

U.S. Bureau of Land Management, 1987, *Washakie resource area resource management plan, portions of Big Horn,* Hot Springs, and Washakie Counties, Wyoming: U.S. Bureau of Land Management, Report no. 870395.

U.S. Bureau of Land Management, 1988, *Challis resource area proposed resource management plan,* upper Columbia-Salmon Clearwater Districts, Custer and Lemhi Counties, Idaho: U.S. Bureau of Land Management, Report no. 980508.

U.S. Bureau of Land Management, 1988, *Cody resource area resource management plan,* Big Horn and Park Counties, Wyoming: U.S. Bureau of Land Management, Report no. 880315.

U.S. Bureau of Land Management, 1988, *Fort Greely national maneuver area and Fort Greely air drop zone,* Alaska—Resource management plan: U.S. Bureau of Land Management, Report no. 880276, 138 p.

U.S. Bureau of Land Management, 1988, *Fort Wainwright maneuver area,* Fairbanks North Star Borough, Alaska: U.S. Bureau of Land Management, Report no. 880277, 138 p.

U.S. Bureau of Land Management, 1988, *Pony Express resource management plan,* Toole, Utah, and Salt Lake Counties, Utah: U.S. Bureau of Land Management, Report no. 880310, 147 p.

U.S. Bureau of Land Management, 1988, *Uncompahgre Basin resource management plan,* Colorado: U.S. Bureau of Land Management, Report no. 880345.

U.S. Bureau of Land Management, 1988, *West Hiline resource management plan,* northcentral Montana: U.S. Bureau of Land Management, Report no. 880200, 306 p.

U.S. Bureau of Land Management, 1988, *Wilderness recommendations for the Red Mountain Wilderness study area in the Arcata resource area,* California: U.S. Bureau of Land Management, Report no. 880033, 120 p.

U.S. Bureau of Land Management, 1989, *Dixie resource area management plan, Washington County,* Utah: U.S. Bureau of Land Management, Report no. 890290, 250 p.

U.S. Bureau of Land Management, 1989, *San Rafael resource management plan,* Emery County, Utah: U.S. Bureau of Land Management, Report no. 890240.

U.S. Bureau of Land Management, 1989, *White Sands resource management plan Amendment for McGregor Range,* Otero County, New Mexico: U.S. Bureau of Land Management, Report no. 890127.

U.S. Bureau of Land Management, 1991, *Safford District resource management plan, Arizona: U.S. Bureau of Land Management,* Report no. 910314, 511 p.

U.S. Bureau of Land Management, 1991, *Three Rivers resource management plan,* Harney, Grant, Lake, and Malheur Counties, Oregon: U.S. Bureau of Land Management, Report no. 910335.

U.S. Bureau of Land Management, 1993, *resource management plan and environmental impact statement for the Big Dry resource area,* Miles City District, Montana: U.S. Bureau of Land Management, Report no. 930076, 382 p.

U.S. Bureau of Land Management, 1994, Fort Wainwright, Yukon maneuver area, *proposed resource management plan and final environmental impact statement,* Fairbanks North Star Borough, Alaska: U.S. Bureau of Land Management, Report no. 940001, 124 p.

U.S. Bureau of Land Management, 1994, *resource management plan and environmental impact statement for the Grass Creek resource area,* Worland District, Wyoming: U.S. Bureau of Land Management, Report, 299 p.

U.S. Bureau of Land Management, 1998, *Judith-Valley-Phillips resource management plan*—Chouteau, Fergus, Judith Basin, Petroleum, Phillips, and Valley Counties, Montana (draft supplement to the final environmental impact statement of September 1997): U.S. Bureau of Land Management, Report no. 980149, 51 p.

U.S. Bureau of Land Management, 1998, *Socorro resource area resource management plan, New Mexico: U.S. Bureau of Land Management, Report no. 880397, 266 p.*

U.S. Bureau of Land Management, 2000, *Federal fluid minerals leasing and development,* Otero and Sierra Counties, New Mexico: U.S. Bureau of Land Management, Report no. 000383, 521 p.

U.S. Bureau of Land Management, 2001, *National management strategy for motorized off-highway vehicle use on public lands:* U.S. Bureau of Land Management, Report no. PB2001103 162, 58 p.

U.S. Bureau of Land Management, 2003, *Farmington resource management plan,* San Juan, McKinley, Rio Arriba, and Sandoval Counties, New Mexico: U.S. Bureau of Land Management, Report no. 030143.

U.S. Bureau of Land Management, 2004, *Dillon resource management plan, Beaverhead and Madison Counties,* Montana: U.S. Bureau of Land Management, Report no. 040153, 626 p.

U.S. Bureau of Land Management, and U.S. Forest Service, 1986, *Proposed Coronado National Forest land and resource management plan,* Arizona and New Mexico: U.S. Bureau of Land Management, and U.S. Department of Agriculture Forest Service, Report no. 860307.

U.S. Bureau of Reclamation, 2001, *Pothole Reservoir resource management plan,* Grant County, Washington: U.S. Bureau of Reclamation, Report no. 010518, 567 p.

U.S. Federal Highway Administration, 1965, *Design of roadside drainage channels:* U.S. Federal Highway Administration, Report no. FHWAEPD86 103, 64 p.

U.S. Fish and Wildlife Service, 1986, *Great Dismal Swamp National Wildlife Refuge master plan,* Virginia and North Carolina: U.S. Fish and Wildlife Service, Report no. 860501, 245 p.

U.S. Forest Service, 1973, *Management of South Holston unit:* U.S. Forest Service, Report no. USDAFSFESADM7350, 121 p.

U.S. Forest Service, 1973, *Proposed off-road vehicle regulations and administrative instructions*: U.S. Forest Service, Report no. USDAFS–FES(ADM)–73–49, 110 p.

U.S. Forest Service, 1973, *Proposed regulations and administrative instructions relating to use of off-road vehicles on National Forest lands:* U.S. Forest Service, Report no. USDAFS–DES(ADM)–73–49, 22 p.

U.S. Forest Service, 1973, *Unit plan for management of the South Holston unit:* U.S. Forest Service, Report no. USDAFS–DES(ADM)–73–50, 80 p.

U.S. Forest Service, 1985, *Caribou National Forest and Curlew National Grassland land and resource management plan,* Idaho, Utah, and Wyoming: U.S. Forest Service, Report no. 850437.

U.S. Forest Service, 1986, *Chequamegon National Forest, Wisconsin land and resource management plan:* U.S. Forest Service, Report no. 860325.

U.S. Forest Service, 1986, *Cleveland National Forest land and resource management plan,* Orange, Riverside, and San Diego Counties, California: U.S. Forest Service, Report no. 860209.

U.S. Forest Service, 1986, *Croatan and Uwharrie National Forests land and resource management plan,* North Carolina: U.S. Forest Service, Report no. 860218.

U.S. Forest Service, 1986, *National Forests in Alabama land and resource management plan:* U.S. Forest Service, Report no. 860095.

U.S. Forest Service, 1986, *Proposed land and resource management plan for Wayne National Forest,* Ohio: U.S. Forest Service, Report no. 860426.

U.S. Forest Service, 1986, *Proposed Santa Fe National Forest land and resource management plan,* Mora, San Miguel, Santa Fe, Sandoval, Los Alamos, and Rio Arriba Counties, New Mexico: U.S. Forest Service, Report no. 860013.

U.S. Forest Service, 1987, *Angeles National Forest land and resource management plan,* Los Angeles, Ventura, and San Bernardino Counties, California: U.S. Forest Service, Report no. 870394.

U.S. Forest Service, 1987, *Proposed Apache-Sitgreaves National Forests land and resource management plan,* Apache, Coconino, Greenlee, and Navajo Counties, Arizona: U.S. Forest Service, Report no. 870387.

U.S. Forest Service, 1987, *Santa Fe National Forest land management plan,* Mora, San Miguel, Santa Fe, Sandoval, Los Alamos, and Rio Arriba Counties, New Mexico: U.S. Forest Service, Report no. 870299.

U.S. Forest Service, 1988, *Proposed Kaibab National Forest land and resource management plan,* Coconino, Yavapai, and Mohave Counties, Arizona: U.S. Forest Service, Report no. 880109.

U.S. Forest Service, 1988, *Record of decision for USDA, Forest Service—Final environmental impact statement,* Allegheny National Forest, Land and resource management plan—Elk, Forest, McKean, and Warren Counties, Pennsylvania: U.S. Forest Service, Technical Report, 41 p.

U.S. Forest Service, 1989, *San Bernardino National Forest land and resource management plan,* San Bernardino and Riverside Counties, California: U.S. Forest Service, Report no. 890023.

U.S. Forest Service, 1990, *Cherokee National Forest land and resource management plan,* Tennessee, North Carolina, and Virginia (final supplement to the final environmental impact statement of April 1986): U.S. Forest Service, Report no. 900062, 62 p.

U.S. Forest Service, 1990, *Umatilla National Forest land and resource management plan,* Oregon and Washington: U.S. Forest Service, Report no. 900237.

U.S. Forest Service, 1992, *Rangeland ecosystem management in the Uinta National Forest,* Provo, Utah: U.S. Forest Service, Report no. 920322, 189 p.

U.S. Forest Service, 1993, *Six Rivers National Forest plan,* Humboldt, Del Norte, Siskiyou, and Trinity Counties, California: U.S. Forest Service, Report no. 930340.

U.S. Forest Service, 1994, *Revised land and resource management plan for the National Forests and Grasslands in Texas*—Angelina, Fannin, Houston, Jasper, Montague, Montgomery, Nacogdoches, Newton, Sabine, San Augustine, San Jacinto, Shelby, Trinity, Walker, and Wise Counties, Texas: U.S. Forest Service, Report no. 940385.

U.S. Forest Service, 1995, *Draft environmental impact statement, Fish Bate analysis area—North Fork Ranger District,* Clearwater National Forest, Clearwater County, Idaho: U.S. Forest Service, Technical Report.

U.S. Forest Service, 1995, *Snowy Trail re-route,* Ventura County, California: U.S. Forest Service, Report no. 950117, 109 p.

U.S. Forest Service, 1996, *Snowy Trail re-route,* Los Padres National Forest, Ventura County, California: U.S. Forest Service, Report no. 960524.

U.S. Forest Service, 2000, *Silver Creek integrated resource project,* Emmett Ranger District, Boise National Forest, Boise and Valley Counties, Idaho: U.S. Forest Service, Report no. 000241, 264 p.

U.S. Forest Service, 2001, *Land and resource management plan,* Uinta National Forest—Juab, Sanpete, Toole, Utah, and Wasatch Counties, Utah: U.S. Forest Service, Report no. 010142.

U.S. Forest Service, 2001, *Starbucky restoration project, Red River Ranger District, Nez Perce National Forest,* Idaho County, Idaho: U.S. Forest Service, Report no. 010274.

U.S. Forest Service, 2002, *Sixshooter project, Emmett Ranger District, Boise National Forest, Gem County,* Idaho: U.S. Forest Service, Report no. 020238, 201 p.

U.S. Forest Service, 2002, *Uncompahgre National Forest travel plan revision,* Gunnison, Hinsdale, Mesa, Montrose, Ouray, San Juan, and San Miguel Counties, Colorado (final supplement to the final environmental impact statement of April 2000): U.S. Forest Service, Report no. 020131, 92 p.

U.S. Forest Service, 2002, *West Gold project, Sandpoint Ranger District,* Idaho Panhandle National Forests, Bonner County, Idaho: U.S. Forest Service, Report no. 020484.

U.S. Forest Service, 2003, *Big Bend Ridge vegetation management project and timber sale,* Ashton/Island Park Ranger District, Caribou-Targhee National Forest, Freemont County, Idaho: U.S. Forest Service, Report no. 030225.

U.S. Forest Service, 2003, *Cross-country travel by off-highway vehicles,* Apache-Sitgreaves, Coconino, Kaibab, Prescott, and Tonto National Forests, Arizona: U.S. Forest Service, Report no. 030184, 235 p.

U.S. Forest Service, 2003, *Duck Creek-Swains access management project,* Cedar City Ranger District, Dixie National Forest, Iron, Garfield, and Kane Counties, Utah: U.S. Forest Service, Report no. 030343, 49 p.

U.S. Forest Service, 2003, *Upper Bear timber sale, Council Ranger District,* Payette National Forest, Adams County, Idaho: U.S. Forest Service, Report no. 030319, 478 p.

U.S. Forest Service, 2004, *Duck Creek fuels treatment analysis,* Cedar City Ranger District, Dixie National Forest, Kane County, Utah: U.S. Forest Service, Report no. 040535, 240 p.

U.S. Forest Service, 2004, *French Face ecosystem restoration,* Ninemile Ranger District, Lolo National Forest, Montana: U.S. Forest Service, Report no. 040336, 184 p.

U.S. Forest Service, 2005, *Dean project area, Bearlodge Ranger District, Black Hills National Forest,* Crook County, Wyoming: U.S. Forest Service, Report no. 050212, 192 p.

U.S. Forest Service, 2005, *Final environmental impact statement for the access designation on the Ocala National Forest,* Lake, Marion, and Putnam Counties, Florida: U.S. Forest Service, Technical Report.

U.S. Forest Service, 2005, *Gallatin National Forest travel management plan,* Gallatin, Madison, Park, Meagher, Sweetgrass, and Carbon Counties, Montana: U.S. Forest Service, Report no. 050231.

U.S. Forest Service, 2005, *Rocky Mountain Ranger District travel management plan,* Rocky Mountain Ranger District, Lewis and Clark National Forest, Glacier, Pondera, Teton, and Lewis and Clark Counties, Montana: U.S. Forest Service, Report no. 050236, 386 p.

U.S. Forest Service, 2006, *Hoosier National Forest land and resource management plan—* Final environmental impact statement: U.S. Forest Service, Technical Report.

U.S. Forest Service, U.S. *Bureau of Land Management, and U.S. Fish and Wildlife Service, 1997,* George Washington National Forest land and resource management plan, Virginia and West Virginia (final supplement to the Final environmental impact statement of January 1993): U.S. Forest Service, U.S. Bureau of Land Management, and U.S. Fish and Wildlife Service, Report no. 970046.

U.S. National Park Service, 1979, *Proceedings of the conference on scientific research in the National Parks,* San Francisco, California, November 26–30, 1979: Washington D.C., U.S. Forest Service.

U.S. National Park Service, 1998, *Redwood National and State Parks general management plan,* Del Norte and Humboldt Counties, California: U.S. National Park Service, Report no. 980290, 474 p.

U.S. Naval Facilities Engineering Command, 2003, *Advanced amphibious assault vehicle, Marine Corps Base Camp Pendleton and San Clemente Range Complex,* San Diego County, California: U.S. Naval Facilities Engineering Command, Report no. 030201.

U.S. Soil Conservation Service, 1990, Erosion, Grand Traverse County, Michigan: *U.S. Soil Conservation Service,* Technical Report.

Valentin, C., Poesen, J., and Li, Y., 2005, *Gully erosion—Impacts, factors and control:* Catena, v. 63, no. 2–3, p. 132–153.

van Der Puy, M.E., 1986, *Rating erosion susceptibility, Proceedings of the fourth federal interagency sedimentation conference,* Las Vegas, Nevada: Minneapolis, Minnesota, Federal Interagency Sedimentation Project, p. 2–1 to 2–7.

Walker, D.A., Webber, P.J., Everett, K.R., and Brown, J., 1977, *The effects of low-pressure wheeled vehicles on plant communities and soils at Prudhoe Bay,* Alaska: Hanover, New Hampshire, Cold Regions Research and Engineering Lab, Report no. CRREL–SR–77–17, 50 p.

Watson, J.J., Kerley, G.I.H., and McLachlan, A., 1996, *Human activity and potential impacts on dune breeding birds in the Alexandria Coastal Dunefield:* Landscape and Urban Planning, v. 34, no. 3–4, p. 3 15–322.

Webb, R.H., 1982, *Off-road motorcycle effects on a desert soil: Environmental Conservation,* v. 9, no. 3, p. 197–208.

Webb, R.H., 1983, *Compaction of desert soils by off-road vehicles, in* Webb, R.H., and Wilshire, H.G., eds., Environmental effects of off-road vehicles—Impacts and management in arid regions: New York, Springer-Verlag, p. 50–79.

Webb, R.H., 2002, *Recovery of severely compacted soils in the Mojave Desert,* California, USA: Arid Land Research and Management, v. 16, no. 3, p. 29 1–305.

Webb, R.H., Ragland, H.C., Godwin, W.H., and Jenkins, D., 1978, *Environmental effects of soil property changes with off road vehicle use: Environmental Management,* v. 2, no. 3, p. 219–233.

Webb, R.H., and Wilshire, H.G., 1983, *Environmental effects of off-road vehicles—Impacts and management in arid regions:* New York, Springer-Verlag, 534 p.

Webb, R.H., Wilshire, H.G., and Henry, M.A., 1983, *Natural recovery of soils and vegetation following human disturbance,* Environmental effects of off-road vehicles-impacts and management in arid regions: New York, Springer-Verlag, p. 279–302.

Wemple, B.C., Jones, J., and Grant, G., 1996, *Channel network extension by logging roads in two basins,* western Cascades, Oregon: Water Resources Bulletin, v. 32, no. 6, p. 1195–1207.

Wester, L., 1994, *Weed management and the habitat protection of rare species—A case study of the endemic Hawaiian fern Marsilea villosa:* Biological Conservation, v. 68, no. 1, p. 1–9.

Wiedemann, A.M., 1984, *Ecology of Pacific Northwest coastal sand dunes—A community profile:* Evergreen State College, Olympia, Washington, Report no. FWSOBS8404, 146 p.

Wilshire, H.G., 1977, *Study results of 9 sites used by off-road vehicles that illustrate land modifications:* U.S. Geological Survey, Technical Report, 22 p.

Wilshire, H.G., 1980, *Human causes of accelerated wind erosion in California's deserts, in Coates,* D.R., and Vitek, J.D., eds., Thresholds in geomorphology: London, George Allen and Unwin, p. 415–433.

Wilshire, H.G., 1983, *Off-road vehicle recreation management policy for public lands in the United States—A case history: Environmental Management,* v. 7, no. 6, p. 489–499.

Wilshire, H.G., 1983, *The impact of vehicles on desert soil stabilizers, in* Webb, R.H., and Wilshire, H.G., eds., Environmental effects of off-road vehicles-impacts and management in arid regions: New York, Springer-Verlag, p. 31–50.

Wilshire, H.G., and Nakata, J.K., 1976, *Off-road vehicle effects on California's Mojave Desert: California Geology,* June 1976, p. 123–132.

Wilshire, H.G., Nakata, J.K., Shipley, Susan, and Prestegaard, Karen, 1978, *Impacts of vehicles on natural terrain at seven sites in the San Francisco Bay area: Environmental Geology,* v. 2, no. 5, p. 295–3 19.

Wilshire, H.G., Shipley, Susan, and Nakata, J.K., 1978, *Impacts of off-road vehicles on vegetation: Transactions of the North American Wildlife and Natural Resources Conference,* v. 43, p. 131–139.

Wilson, J.P., and Seney, J.P., 1994, *Erosional impact of hikers, horses, motorcycles, and off-road bicycles on mountain trails in Montana: Mountain Research and Development,* v. 14, no. 1, p. 77–88.

Wolcott, T.G., and Wolcott, D.L., 1984, *Impact of off-road vehicles on macroinvertebrates of a mid-Atlantic beach: Biological Conservation,* v. 29, no. 3, p. 2 17–240.

Wright, R.C., and Burns, J.R., 1968, *Mobility environmental research study—A quantitative method for describing terrain for ground mobility.* Volume 2. Surface compound: Vicksburg, Mississippi, U.S. Army Engineer Waterways Experiment Station, Report no. AWES–TR–3–726–2, 99 p.

Zander, J., Ammer, U., and Breitsameter, J., 1988, *Experimental studies on the effect of off-road traffic on topsoil compaction: Forstwissenschaftliches Centralblatt,* v. 107, no. 2, p. 112–122.

Ziani, F., and Biarez, J., 1990, *Pressure sinkage relationship for tyres on very loose sand: Journal of Terramechanics,* v. 27, no. 3, p. 167–177.

1.1. OHV Effects on Vegetation

Agrawal, Y.K., Patel, M.P., and Merh, S.S., 1981, Lead in soils and plants—Its relationship to traffic volume and proximity to highway (Lalbag, Baroda City): *International Journal of Environmental Studies,* v. 16, no. 3–4, p. 222–224.

Ahlstrand, G.M., and Racine, C.H., 1993, Response of an Alaska, U.S.A., shrub-tussock community to selected all-terrain vehicle use: *Arctic and Alpine Research* v. 25, no. 2, p. 142–149.

Al-Awadhi, J.I., 2001, Impact of gravel quarrying on the desert environment of Kuwait: *Environmental Geology,* v. 41, no. 3–4, p. 365–371.

Albrecht, J., and Knopp, T.B., 1985, *Off road vehicles—Environmental impact—Management response—A bibliography:* Rosemount, Minnesota, Agricultural Experiment Station, University of Minnesota, Miscellaneous Publication, 50 p.

Anders, F.J., and Leatherman, S.P., 1987, *Effects of off-road vehicles on coastal foredunes at Fire Island,* New York, USA: Environmental Management, v. 11, no. 1, p. 45–52.

Angold, P.G., 1997, The impact of a road upon adjacent heathland vegetation—Effects on plants species composition: *Journal of Applied Ecology,* v. 34, no. 2, p. 409–417.

Assistant Secretary of Defense (Health and Environment), 1973, *Proposed Department of Defense regulation on use of off-road vehicles:* U.S. Department of Defense, Report no. ELR0511, 20 p.

Bazzaz, F.A., and Garbutt, K., 1988, The response of annuals in competitive neighborhoods— Effects of elevated CO2: *Ecology,* v. 69, no. 4, p. 937–946.

Beije, H.M., 1986, Effects of military-training activities on soil, vegetation, and fauna: Leersum, Netherlands, *Research Institute for Nature Management,* p. 95–111.

Belnap, Jayne, 1995, Surface disturbances—Their role in accelerating desertification: *Environmental Monitoring and Assessment,* v. 37, no. 1–3, p. 39–57.

Ben Brooks and Associates, Ironhorse Investors Inc, *Santa Fe Pacific Railroad Company and U.S. Department of the Interior,* Bureau of Land Management, 1998, Hualapai Mountain Land Exchange, Mohave County, Arizona, Report no. 980482, 175 p.

Benninger-Traux, M., Vankat, J.L., and Schaefer, R.L., 1992, *Trail corridors as habitat and conduits for movement of plant species in Rocky Mountain National Park,* Colorado, USA: Landscape Ecology, v. 6, no. 4, p. 269–278.

Berry, K.H, 1980, *A review of the effects of off-road vehicles on birds and other vertebrates, in DeGraaf, R.M., and Tilghman,* N.G., eds., Management of western forests and grasslands for nongame birds—Workshop proceedings: Ogden Utah, U.S. Forest Service, Intermountain Forest and Range Experiment Station, p. 451–467.

Berry, K.H., 1980, The effects of four-wheel vehicles on biological resources—Soils, vegetation, small animals, management problems, Southwestern United States, *in* Andrews, R.N.L., and Nowak, P.F., eds., *Off-road vehicle use—A management challenge:* Washington D.C., U.S. Office of Environmental Quality, p. 23 1–233.

Boer, B., 1998, Anthropogenic factors and their potential impacts on the sustainable development of Abu Dhabi's terrestrial biological resources: *International Journal of Sustainable Development and World Ecology,* v. 5, no. 2, p. 125–135.

Bolling, J.D., and Walker, L.R., 2000, Plant and soil recovery along a series of abandoned desert roads: *Journal of Arid Environments,* v. 46, no. 11, p. 1–24.

Brainard, J., 1998, Patton Tank marks suggest long recovery: *Science News*, v. 154, no. 6, p. 87.

Brattstrom, B.H., 2000, The range, habitat requirements, and abundance of the orange-throated whiptail, *Cnemidophorus hyperythrus beldingi*: *Bulletin of the Southern California Academy of Sciences*, v. 99, no. 1, p. 1–24.

Braunack, M.V., 1986, The residual effects of tracked vehicles on soil surface properties: *Journal of Terramechanics,* v. 23, no. 1, p. 37–50.

Brodhead, J.M.B., and Godfrey, P.J., 1977, Off road vehicle impact in Cape Cod National Seashore—Disruption and recovery of dune vegetation: *International Journal of Biometeorology,* v. 21, no. 3, p. 299–306.

Brodhead, J.M.B., 1978, *Monitoring of Province Lands off-road vehicle impact sites: Amherst,* Massachusetts, Massachusetts University and National Park Service Cooperative Research Unit, Report no. UM–NPSCRU–40, 21 p.

Brodhead, J.M.B., 1979, *Remonitoring of province lands off-road vehicle impact sites:* Amherst, Massachusetts, Massachusetts University and National Park Service Cooperative Research Unit, Report no. UM–NPSCRU–43, 10 p.

Brooks, M.L., 1995, *Benefits of protective fencing to plant and rodent communities of the western Mojave Desert,* California: Environmental Management, v. 19, no. 1, p. 65–74.

Broughton, J.D., and Adder, E.E., 1968, *Mobility environmental research study—A quantitative method for describing terrain for ground mobility.* Volume IV. Vegetation: Vicksburg, Mississippi, U.S. Army Engineer Waterways Experiment Station, Report no. AEWES–TR–3–726–4, 159 p.

Brown, G., and Porembski, S., 2000, Phytogenic hillocks and blow-outs as 'safe sites' for plants in an oil-contaminated area of northern Kuwait: *Environmental Conservation,* v. 27, no. 3, p. 242–249.

Brown, G., and Schoknecht, N., 2001, Off-road vehicles and vegetation patterning in a degraded desert ecosystem in Kuwait: *Journal of Arid Environments,* v. 49, no. 2, p. 413–427.

Charman, D.J., and Pollard, A.J., 1995, Long-term vegetation recovery after vehicle track abandonment on Dartmoor, SW England, U.K.: *Journal of Environmental Management,* v. 45, no. 1, p. 73–85.

Clampitt, C.A., 1993, Effects of human disturbances on prairies and the regional endemic *Aster curtus* in western Washington: *Northwest Science,* v. 67, no. 3, p. 163–169.

Collins, E., O'Farrell, T.P., and Rhoads, W., 1982, *Annotated bibliography for biologic overview for the Nevada nuclear waste storage investigations,* Nevada Test Site, Nye County, Nevada: Goleta, California, EG and G, Inc., Report no. EGG1 1832419, 48 p.

Collins, E., O'Farrell, T.P., and Rhoads, W., 1982, *Biologic overview for the Nevada nuclear waste storage investigations,* Nevada Test Site, Nye County, Nevada: Goleta, California, EG and G, Inc., Report no. EGG1 1832460, 55 p.

Davidson, E.D., and Fox, M., 1974, *Effects of off-road motorcycle activity on Mojave Desert vegetation and soil:* Madroño, v. 22, no. 8, p. 381–390.

Duever, M.J., Riopelle, L.A., and McCollom, J.M., 1986, *Long term recovery of experimental off-road vehicle impacts and abandoned old trails in the Big Cypress National Preserve:* Naples, Florida, National Audubon Society, Technical Report, 56 p.

Dyke, L.D., 1985, *Terrain disturbance due to summer off-road vehicle use in central Keewatin,* Northwest Territories, Canada: Ottawa, Ontario, Department of Indian Affairs and Northern Development, Report no. SSC–R71–19/36–1985F, 53 p.

Everett, K.R., Murray, B.M., Murray, D.F., Johnson, A.W., and Linkins, A.E., 1985, *Reconnaissance observations of long-term natural vegetation recovery in the Cape Thompson region,* Alaska, and additions to the checklist of flora: Hanover, New Hampshire, Cold Regions Research and Engineering Lab, Report no. CRREL–85–1 1, 85 p.

Falkengren-Grerup, U., 1986, *Soil acidification and its impact on ground vegetation: Oecologia,* v. 70, p. 339–347.

Fang, S., Wente, S., Gertner, G.Z., Wang, G., and Anderson, A., 2002, Uncertainty analysis of predicted disturbance from off-road vehicular traffic in complex landscapes at Fort Hood: *Environmental Management,* v. 30, no. 2, p. 199–208.

Fatoki, O.S., 2000, Trace zinc and copper concentrations in roadside vegetation and surface soils—A measurement of local atmospheric pollution in Alice, South Africa: International *Journal of Environmental Studies* (UK), v. 57, no. 5, p. 501–5 13.

Felix, N.A., Raynolds, M.K., Jorgenson, J.C., and Dubois, K.E., 1992, Resistance and resilience of tundra plant communities to disturbance by winter seismic vehicles: *Arctic and Alpine Research,* v. 24, no. 1, p. 69–77.

Ferris, R., and Taylor, G., 1995, Contrasting effects of elevated CO2 and water deficit on two native herbs: *New Phytologist,* v. 131, no. 4, p. 491–501.

Frodigh, R.J., 1967, *Seasonal change revealed by time-lapse photography:* Natick, Massachusetts, Army Natick Labs Earth Sciences Division, Report no. USANLABS–TR–67–34–ES, 61 p.

Fuchs, E.H., Wood, M.K., Jones, T.L., and Racher, B., 2003, Impacts of tracked vehicles on sediment from a desert soil: *Journal of Range Management,* v. 56, no. 4, p. 342–3 52.

Godfrey, P.J., Leatherman, S.P., and Buckley, P.A., 1978, Impact of off-road vehicles on coastal ecosystems, Coastal zone '78—*Proceedings of the symposium on technical, environmental, socioeconomic and regulatory aspects of coastal zone management,* San Francisco, California, March 14–16, 1978.

Griggs, G.B., and Walsh, B.L., 1981, The impact, control, and mitigation of off-road vehicle activity in Hungry Valley, *California: Environmental Geology,* v. 3, no. 4, p. 229–243.

Grimes, D.W., Miller, R.J., and Wiley, P.L., 1975, Cotton and corn root development in two field soils of different strength characteristics: *Agronomy Journal,* v. 67, p. 5 19–523.

Grimes, D.W., Sheesley, W.R., and Wiley, P.L., 1978, Alfalfa root development and shoot regrowth in compact soil of wheel traffic patterns: *Agronomy Journal,* v. 70, p. 955–958.

Griswold, T.L., 1996, A new *Microbembex* endemic to the Algodones Dunes, California (Hymenoptera: Sphecidae): *Pan-Pacific Entomologist,* v. 72, no. 3, p. 142–144.

Hall, C., and Dearden, P., eds., 1984, T*he impact of "non-consumptive" recreation on wildlife— An annotated bibliography:* Monticello, Illinois, Vance Bibliographies, 45 p.

Hansen, D.J., and Ostler, W.K., 2001, *Plant-damage assessment technique for evaluating military vehicular impacts to vegetation in the Mojave Desert*: Las Vegas, Nevada, Bechtel Nevada, Inc., Las Vegas, NV, Report no. DOE/NV /1178613.

Hillel, D., and Tadmor, N., 1962, Water regime and vegetation in central Negev Highlands of Israel: *Ecology,* v. 43, no. 1, p. 33–41.

Hirst, R.A., Pywell, R.F., Marrs, R.H., and Putwains, P.D., 2003, The resistance of a chalk grassland to disturbance: *Journal of Applied Ecology,* v. 40, no. 2, p. 3 68–379.

Holzapfel, C., and Schmidt, W., 1990, Roadside vegetation along transects in the Judean Desert: I*srael Journal of Botany*, v. 39, p. 263–270.

Hosier, P.E., and Eaton, T.E., 1980, *The impact of vehicles on dune and grassland vegetation on a south-eastern North Carolina barrier beach:* Wilmington, North Carolina, North Carolina University at Wilmington, Report no. UNC–SG–R–1 64, 13 p.

Hunt, A., Dickens, H.J., and Whelan, R.J., 1987, Movement of mammals through tunnels under railway lines: *Australian Zoologist,* v. 24, no. 2, p. 89–93.

Iverson, R.M., Hinckley, B.S., and Webb, R.M., 1981, Physical effects of vehicular disturbances on arid landscapes: *Science,* v. 212, no. 4497, p. 915–917.

Johnson, R.R., Mills, G.S., and Carothers, S.W., 1990, Creation and restoration of riparian habitat in southwestern arid and semi-arid regions, *in* Kusler, J.A., and Kentula, M.E., eds., Wetland creation and restoration—*The status of the science:* Covelo, California, Island Press, p. 351–366.

Johnson, R.R., and Simpson, J.M., 1988, Desertification of wet riparian ecosystems in arid regions of the North American southwest, *in* Whitehead, E.E., Hutchinson, C.F.,

Timmermann, B.N., and Varady, R.G., eds., *Arid lands—Today and tomorrow:* Tuscon, Arizona, October 20–25, 1985: Boulder, Colorado, Westview Press, p. 1383–1393.

Johnston, F.M., and Johnston, S.W., 2004, Impacts of road disturbance on soil properties and on exotic plant occurrence in subalpine areas of the Australian Alps: *Arctic, Antarctic, and Alpine Research,* v. 36, no. 2, p. 201–207.

Kay, B.L., and Graves, W.L., 1983, Revegetation and stabilization techniques for disturbed desert vegetation, *in* Webb, R.H., and Wilshire, H.G., eds., *Environmental effects of off-road vehicles—Impacts and management in arid regions:* New York, Springer-Verlag, p. 3 25–340.

Kay, J., 1981, Evaluating environmental impacts of off-road vehicles: *Journal of Geography,* v. 80, no. 1, p. 10–18.

Kert, J., 1992, *Remote sensing of off-road vehicle emissions, Proceedings of the optical remote sensing and applications to environmental and industrial safety problems conference,* Houston, Texas, April 6–8, 1992.

Knisley, C.B., and Hill, J.M., 2001, Biology and conservation of the Coral Pink Sand Dunes tiger beetle, *Cicindela limbata albissima* Rumpp: *Western North American Naturalist,* v. 61, no. 4, p. 381–394.

Kondolf, G.M., Piegay, H., and Landon, N., 2002, Channel response to increased and decreased bedload supply from land use change—Contrasts between two catchments: *Geomorphology,* v. 45, no. 1–2, p. 35–51.

Krzysik, A.J., 1994, *Desert tortoise at Fort Irwin, California—A federal threatened species:* Champaign, Illinois, Construction Engineering Research Lab, Report no. CERLEN9410, 102 p.

Kurczewski, F.E., 1998, Distribution, status, evaluation, and recommendations for the protection of *Tachysphex pechumani* Krombein, the antennal-waving wasp: *Natural Areas Journal,* v. 18, no. 3, p. 242–254.

Kurczerski, F.E., 2000, *History of white pine* (*Pinus strobus*)/oak (*Quercus* spp.) savanna in southern Ontario, with particular reference to the biogeography and status of the antenna-waving wasp, *Tachysphex pechumani* (Hymenoptera: Sphecidae): Canadian Field-Naturalist, v. 114, no. 1, p. 1–20.

Kutiel, P., Eden, E., and Zhevelev, Y., 2000, Effect of experimental trampling and off-road motorcycle traffic on soil and vegetation of stabilized coastal dunes, Israel: *Environmental Conservation,* v. 27, no. 1, p. 14–23.

Kutiel, P., Eden, Z., and Zhevelev, H., 2001, The impact of motorcycle traffic on soil and vegetation of stabilized coastal dunes, Israel: *Journal of Coastal Conservation,* v. 7, no. 1, p. 81–89.

Langdon, A.M., 2000, *Mojave Desert soils, plants, and ants—Developing a monitoring strategy for off-highway-vehicles:* Pomona, California, California State Polytechnic University, Pomona, 114 p.

Lathrop, E.W., 1983, The effect of vehicle use on desert vegetation, *in* Webb, R.H., and Wilshire, H.G., eds., *Environmental effects of off-road vehicles—Impacts and management in arid regions:* New York, Springer-Verlag, p. 153–166.

Leis, S.A., Engle, D.M., Leslie, D.M., and Fehmi, J.S., 2005, Effects of short- and long-term disturbance resulting from military maneuvers on vegetation and soils in a mixed prairie area: *Environmental Management,* v. 36, no. 6, p. 849–861.

Lightfoot, D.C., and Whitford, W.G., 1991, Productivity of creosotebush foliage and associated canopy arthropods along a desert roadside: *American Midland Naturalist,* v. 125, no. 2, p. 310–322.

Lovich, J.E., and Bainbridge, D., 1999, Anthropogenic degradation of the southern California desert ecosystem and prospects for natural recovery and restoration: *Environmental Management,* v. 24, no. 3, p. 309–326.

Luckenbach, R.A., 1982, *Ecology and management of the desert tortoise (Gopherus agassizii) in California:* U.S. Fish and Wildlife Service, Technical Report, 37 p.

Meffe, G.K., and Carroll, R.C., 1997, *Principles of conservation biology: Sunderland, Massachusetts,* Sinauer Associates, Inc., 729 p.

Misak, R.F., Al-Awadhi, J.M., Omar, S.A., and Shahid, S.A., 2002, *Soil degradation in Kabd area, southwestern Kuwait City: Land Degradation and Development,* v. 13, no. 5, p. 403–415.

Mize, R., Evans, R.E., Macroberts, B.R., Macroberts, M.H., and Rudolph, D.C., 2005, Restoration of pitcher plant bogs in eastern Texas, USA: *Natural Areas Journal,* v. 25, no. 2, p. 197–201.

Mooney, H.A., Kueppers, M., Koch, G., Gorham, J., Chu, C., and Winner, W.E., 1988, Compensating effects to growth of carbon partitioning changes in response to SO2-induced photosynthetic reduction in radish: *Oecologia,* v. 75, no. 4, p. 502–506.

Motto, H.L., Daines, R.H., Chilko, D.M., and Motto, C.K., 1970, Lead in soils and plants—Its relationship to traffic volume and proximity to highways: *Environmental Science and Technology,* v. 4, p. 231–237.

Munger, J.C., Barnett, B.R., Novak, S.J., and Ames, A.A., 2003, *Impacts of off-highway motorized vehicle trails on the reptiles and vegetation of the Owyhee Front:* Idaho, U.S. Bureau of Land Management Technical Bulletin No. 03-3, *http://www .id.blm.gov/techbuls/03_03/doc.pdf.*

Noren, O., Danfors, B., and Stambeck, A., 1984, *Technical systems in energy forestry:* Stockholm, Sweden, Statens Energiverk, Report no. STEV–EO–84–4, 145 p.

O'Farrell, T.P., 1984, *Conservation of the endangered San Joaquin kit fox Vulpes macrotis mutica in the Naval Petroleum Reserves,* California: Acta Zoologica Fennica, v. 172, p. 207–208.

Onyeanusi, A.E., 1986, Measurements of impact of tourist off-road driving on grasslands in Masai Mara National Reserve, Kenya—A simulation approach: *Environmental Conservation,* v. 13, no. 4, p. 325–329.

Parendes, L.A., and Jones, J.A., 2000, Role of light availability and dispersal in exotic plant invasion along roads and streams in the H. J. Andrews Experimental Forest, Oregon: *Conservation Biology,* v. 14, no. 1, p. 64–75.

Persico, L.P., Nichols, K.K., and Bierman, P.R., 2005, Tracking painted pebbles—Short-term rates of sediment movement on four Mojave Desert piedmont surfaces: *Water Resources Research,* v. 41, no. 7, p. W07004.07001–W07004.07015.

Pinard, M.A., Barker, M.G., and Tay, J., 2000, Soil disturbance and post-logging forest recovery on bulldozer paths in Sabah, Malaysia: *Forest Ecology and Management,* v. 130, no. 1–3, p. 213–225.

Prose, D.V., Metzger, S.K., and Wilshire, H.G., 1987, Effects of substrate disturbance on secondary plant succession—Mojave Desert, California: *Journal of Applied Ecology,* v. 24, no. 1, p. 305–313.

Quarles, H.D., Hanawalt, R.B., and Odum, W.E., 1974, Lead in small mammals, plants, and soil at varying distances from a highway: *Journal of Applied Ecology,* v. 11, no. 3, p. 937–949.

Rickard, C.A., McLachlan, A., and Kerley, G.I.H., 1994, The effects of vehicular and pedestrian traffic on dune vegetation in South Africa: *Ocean and Coastal Management,* v. 23, no. 3, p. 225–247.

Rickard, W.H., 1988, *Natural vegetation at the proposed reference repository location in southeastern Washington:* Richland, Washington, Battelle Pacific Northwest Labs, Report no. PNL–6402, 27 p.

Schemnitz, S.D., and Schortemeyer, J.D., 1972, T*he influence of vehicles on Florida Everglades vegetation:* Tallahassee, Florida, Florida State Game and Fresh Water Fish Commission, Report no. DISFEP–74–3 1, 65 p.

Schreiner, B.G., 1963, Speed tests conducted in Canada during Muskeg Trafficability Test Program, August 1962: Vicksburg, Mississippi, *U.S. Army Engineer Waterways Experiment Station,* Report no. AEWES–MISC–PAPER–4–621, 26 p.

Schultink, G., 1977, *Impact analysis of off-road-vehicle use on vegetation in the Grand Mere Dune environment:* East Lansing, Michigan, Michigan State University, Report no. NASACR155764, 10 p.

Shay, R., 1979, Management problems in off-road-vehicle recreation, *in* Ittner, R., Potter, D.R., Agee, J.K., and Anschell, S., eds., *Recreational impact on wildlands—Conference proceedings, Seattle, Washington, October 27–29, 1978,* U.S. Forest Service, Pacific Northwest Forest and Range Experiment Station and U.S. National Park Service, p. 314–317.

Sheridan, D., 1979, *Off-road vehicles on public land:* Washington, D.C., Council on Environmental Quality, Report no. PB8621 1158, 94 p.

Shoop, S.A., 1993, Terrain characterization for trafficability: Hanover, New Hampshire, *Cold Regions Research and Engineering Lab,* Report no. CRREL–93–6, 29 p.

Smith, W.H., 1976, Lead contamination of the roadside ecosystem: *Journal of the Air Pollution Control Association, v.* 26, p. 753–766.

Snyder, C.T., Frickel, D.G., Hadley, R.F., and Miller, R.F., 1976, Effects of off-road vehicle use on the hydrology and landscape of arid environments in central and southern California: *U.S. Geological Survey, Water Resources Investigations Report* USGS/WRI–76–99, 52 p.

Sparrow, S.D., Wooding, F.J., and Whiting, E.H., 1978, Effects of off-road vehicle traffic on soils and vegetation in the Denali Highway region of Alaska: *Journal of Soil and Water Conservation,* v. 33, no. 1, p. 20–27.

Spellerburg, I.F., and Morrison, T., 1998, *The ecological effects of new roads—A literature review:* Wellington, New Zealand, New Zealand Department of Conservation, Technical Report, 55 p.

Spencer, H.J., and Port, G.R., 1988, Effects of roadside conditions on plants and insects: II. Soil conditions: *Journal of Applied Ecology,* v. 25, no. 22, p. 709–7 15.

Stanton, E.J., and Kurczewski, F.E., 1999, Notes on the distribution of *Cicindela lepida* Dejean (Coleoptera: Cicindelidae) in New York, Ontario and Quebec: *Coleopterists Bulletin,* v. 53, no. 3, p. 275–279.

Steiner, A.J., and Leatherman, S.P., 1979, *An annotated bibliography of the effects of off-road vehicle and pedestrian traffic on coastal ecosystems:* U.S. Department of Agriculture,

U.S. National Park Service Cooperative Research Unit, and University of Massachusetts Amherst, Report no. UMNPSCRU45, 88 p.

Stinson, B., 1971, An analytical model for predicting cross-country vehicle performance. *Appendix E-Quantification of the screening effects of vegetation on driver's vision and vehicle speed*: Vicksburg, Mississippi, U.S. Army Engineer Waterways Experiment Station, Report no. AEWESTR3783, 33 p.

Thompson, C.R., Olszyk, D.M., Kats, G., Bytnerowicz, A., Dawson, P.J., and Wolf, J.W., 1984, Effects of ozone or sulfur dioxide on annual plants of the Mojave Desert: *Journal of the Air Pollution Control Association,* v. 34, no. 10, p. 1017–1022.

Tunstall, B.R., and Reece, P.H., 1989, *Environmental assessment of the Sunset and Big Desert lands, northwest Victoria,* Australia: Clayton South, Australia, Australia Commonwealth Scientific and Industrial Research Organization, Division of Water Resources Divisional Report, Technical Report, p. 1–85.

Turton, S.M., 2005, Managing environmental impacts of recreation and tourism in rainforests of the wet tropics of Queensland World Heritage Area: *Geographical Research,* v. 43, no. 2, p. 140–151.

Tuttle, M., and Griggs, G., 1987, Soil erosion and management recommendations at three State Vehicular Recreation Areas, California: *Environmental Geology and Water Sciences,* v. 10, no. 2, p. 111–123.

Tyser, R.W., and Worley, C.A., 1992, Alien flora in grasslands adjacent to road and trail corridors in Glacier National Park, Montana (U.S.A.). *Conservation Biology,* v. 6, no. 2, p. 253-262.

U.S. Army Corps of Engineers, 1994, *Stationing of mechanized or armored combat forces at Fort Lewis,* Thurston and Pierce Counties, Washington: U.S. Army Corps of Engineers, Report no. 940045.

U.S. Army Corps of Engineers, 1998, *McGregor range land withdrawal, Fort Bliss, Otera County,* New Mexico: U.S. Army Corps of Engineers, Report no. 980439, 586 p.

U.S. Army Corps of Engineers, 2001, *Fort Bliss mission and master plan—Dona Ana and Otero Counties,* New Mexico, and El Paso County, Texas: U.S. Army Corps of Engineers, Report no. 010081.

U.S. Army Corps of Engineers, and Oklahoma Department of Wildlife Conservation, 1986, Candy Lake, Candy Creek, Osage County, Oklahoma (supplemental information report to the final environmental impact statement of January 1975): U.S. Army Corps of Engineers, Report no. 860246, 4 p.

U.S. Bureau of Land Management, 1985, *Box Elder resource management plan,* Utah: U.S. Bureau of Land Management, Report no. 850448.

U.S. Bureau of Land Management, 1985, *Lower Gila South resource management plan, La Paz, Maricopa, Pima, Pinal, and Yuma Counties, Arizona:* U.S. Bureau of Land Management, Report no. 850347, 909 p.

U.S. Bureau of Land Management, 1985, *Proposed Two Rivers resource management plan,* Oregon: U.S. Bureau of Land Management, Report no. 850409.

U.S. Bureau of Land Management, 1985, *Resource management plan for the Kemmerer resource area,* Lincoln, Sublette, Sweetwater, and Uinta Counties, Wyoming: U.S. Bureau of Land Management, Report no. 850532.

U.S. Bureau of Land Management, 1985, *Resource management plan for the Walker Planning Area,* Nevada: U.S. Bureau of Land Management, Report no. 850274.

U.S. Bureau of Land Management, 1985, *Rio Puerco resource management plan,* New Mexico: U.S. Bureau of Land Management, Report no. 850417.

U.S. Bureau of Land Management, 1985, *Two Rivers resource management plan*, Oregon: U.S. Bureau of Land Management, Report no. 850137, 163 p.

U.S. Bureau of Land Management, 1985, *White Sands resource area management plan*, New Mexico: U.S. Bureau of Land Management, Report no. 850387.

U.S. Bureau of Land Management, 1985, *Yuma District resource management plan,* Yuma, La Paz, and Mohave Counties, Arizona, and San Bernardino, Riverside, and Imperial Counties, California: U.S. Bureau of Land Management, Report no. 850358, 310 p.

U.S. Bureau of Land Management, 1986, *Baker resource management plan,* Baker County, Oregon: U.S. Bureau of Land Management, Report no. 860150, 129 p.

U.S. Bureau of Land Management, 1986, *Lander resource management plan,* Lander, Wyoming: U.S. Bureau of Land Management, Report no. 860449.

U.S. Bureau of Land Management, 1986, *Proposed 1985 amendments to the California desert conservation area plan and the Eastern San Diego County Master Framework Plan,* California: U.S. Bureau of Land Management, Report no. 860432, 138 p.

U.S. Bureau of Land Management, 1986, *Proposed 1985 amendments to the California desert conservation area plan,* California: U.S. Bureau of Land Management, Report no. 860077.

U.S. Bureau of Land Management, 1986, *Proposed Baker resource management plan,* Vale District, Oregon: U.S. Bureau of Land Management, Report no. 860396.

U.S. Bureau of Land Management, 1986, *resource management plan for the Carlsbad resource area,* Eddy, Lea, and Chaves Counties, New Mexico: U.S. Bureau of Land Management, Report no. 860101, 390 p.

U.S. Bureau of Land Management, 1986, *resource management plan for the House Range resource area,* Juab and Millard Counties, Utah: U.S. Bureau of Land Management, Report no. 860377.

U.S. Bureau of Land Management, 1986, *resource management plan for the Warm Springs resource area,* Millard County, Utah: U.S. Bureau of Land Management, Report no. 860395.

U.S. Bureau of Land Management, 1986, *Shoshone/Sun Valley Wilderness Study, Blaine, Butte, Cama, Custer, Gooding, and Lincoln Counties,* Idaho: U.S. Bureau of Land Management, Report no. 860148, 225 p.

U.S. Bureau of Land Management, 1987, *Cascade resource management plan, Idaho:* U.S. Bureau of Land Management, Report no. 870281, 432 p.

U.S. Bureau of Land Management, 1987, *North Dakota resource management plan, Dunn and Bowman Counties*, North Dakota: U.S. Bureau of Land Management, Report no. 870236, 199 p.

U.S. Bureau of Land Management, 1987, *Pinedale resource area resource management plan,* Sublette and Lincoln Counties, Wyoming: U.S. Bureau of Land Management, Report no. 870077, 347 p.

U.S. Bureau of Land Management, 1987, *Pinedale resource area resource management plan,* Sublette, Teton, Lincoln, and Fremont Counties, Wyoming: U.S. Bureau of Land Management, Report no. 870428, 232 p.

U.S. Bureau of Land Management, 1987, *Pocatello resource management plan,* Idaho: U.S. Bureau of Land Management, Report no. 870333, 132 p.

U.S. Bureau of Land Management, 1987, *Preliminary wilderness recommendations for the Arcata resource area,* Eden Valley and Thatcher Ridge Wilderness Study Areas, Ukiah District, Mendocino County, California: U.S. Bureau of Land Management, Report no. 870348, 111 p.

U.S. Bureau of Land Management, 1987, *Proposed Monument resource management plan and Wilderness Study,* Idaho: U.S. Bureau of Land Management, Report no. 870248, 162 p.

U.S. Bureau of Land Management, 1987, *Proposed resource management plan for the San Juan resource area,* Moab District, Utah: U.S. Bureau of Land Management, Report no. 870440.

U.S. Bureau of Land Management, 1987, *Proposed resource management plan and wilderness recommendations for the Lahontan resource area,* Nevada: U.S. Bureau of Land Management, Report no. 870293, 282 p.

U.S. Bureau of Land Management, 1987, *Proposed utility corridor resource management plan, Alaska: U.S. Bureau of Land Management,* Report no. 870291, 281 p.

U.S. Bureau of Land Management, 1987, *resource management plan for the Shoshone—Eureka resource area, Nevada—Wilderness recommendations:* U.S. Bureau of Land Management, Report no. 870403, 101 p.

U.S. Bureau of Land Management, 1987, *Taos resource management plan,* New Mexico: U.S. Bureau of Land Management, Report no. 870331.

U.S. Bureau of Land Management, 1987, *Washakie resource area resource management plan,* portions of Big Horn, Hot Springs, and Washakie Counties, Wyoming: U.S. Bureau of Land Management, Report no. 870395.

U.S. Bureau of Land Management, 1987, *Winnemucca District wilderness recommendations,* Nevada and California: U.S. Bureau of Land Management, Report no. 870342, 581 p.

U.S. Bureau of Land Management, 1988, *Challis resource area proposed resource management plan,* upper Columbia-Salmon Clearwater Districts, Custer and Lemhi Counties, Idaho: U.S. Bureau of Land Management, Report no. 980508.

U.S. Bureau of Land Management, 1988, *Cody resource area resource management plan,* Big Horn and Park Counties, Wyoming: U.S. Bureau of Land Management, Report no. 880315.

U.S. Bureau of Land Management, 1988, *Fort Greely national maneuver area and Fort Greely air drop zone,* Alaska—Resource management plan: U.S. Bureau of Land Management, Report no. 880276, 138 p.

U.S. Bureau of Land Management, 1988, *Fort Wainwright maneuver area, Fairbanks North Star Borough,* Alaska: U.S. Bureau of Land Management, Report no. 880277, 138 p.

U.S. Bureau of Land Management, 1988, *Medicine Bow-Divide resource areas (now Great Divide resource area) resource management plan,* Rawlins District, Wyoming: U.S. Bureau of Land Management, Report no. 880325.

U.S. Bureau of Land Management, 1988, *Proposed Las Vegas resource management plan and final environmental impact statement,* Clark and Nye Counties, Nevada: U.S. Bureau of Land Management, Report no. 980207.

U.S. Bureau of Land Management, 1988, *Uncompahgre Basin resource management plan,* Colorado: U.S. Bureau of Land Management, Report no. 880345.

U.S. Bureau of Land Management, 1988, *Wilderness recommendations for the Red Mountain Wilderness study area in the Arcata resource area,* California: U.S. Bureau of Land Management, Report no. 880033, 120 p.

U.S. Bureau of Land Management, 1989, *Dixie resource area management plan, Washington County,* Utah: U.S. Bureau of Land Management, Report no. 890290, 250 p.

U.S. Bureau of Land Management, 1989, *Phoenix resource management plan, Arizona: U.S. Bureau of Land Management,* Report no. 890012, 240 p.

U.S. Bureau of Land Management, 1989, *Powder River Wilderness suitability study for the Powder River resource area resource management plan,* Miles City District, Montana: U.S. Bureau of Land Management, Report no. 890301, 73 p.

U.S. Bureau of Land Management, 1989, *San Pedro River riparian management plan for the San Pedro River EIS Area,* Cochise County, Arizona: U.S. Bureau of Land Management, Report no. 890152, 381 p.

U.S. Bureau of Land Management, 1989, *San Rafael resource management plan, Emery County,* Utah: U.S. Bureau of Land Management, Report no. 890240.

U.S. Bureau of Land Management, 1989, *Utility corridor proposed resource management plan,* Alaska: U.S. Bureau of Land Management, Report no. 890326, 281 p.

U.S. Bureau of Land Management, 1989, *White Sands resource management plan amendment for McGregor Range,* Otero County, New Mexico: U.S. Bureau of Land Management, Report no. 890127.

U.S. Bureau of Land Management, 1990, *Bishop resource management plan and environmental impact statement:* U.S. Bureau of Land Management, Report no. BLMCAES90001 1610, 273 p.

U.S. Bureau of Land Management, 1991, *Arizona Strip District resource management plan, Mohave and Coconino Counties,* Arizona: U.S. Bureau of Land Management, Report no. 910055, 617 p.

U.S. Bureau of Land Management, 1991, *Safford District resource management plan, Arizona: U.S. Bureau of Land Management,* Report no. 910314, 511 p.

U.S. Bureau of Land Management, 1991, *Spokane resource management plan, Washington: U.S. Bureau of Land Management*, Report no. 910374, 169 p.

U.S. Bureau of Land Management, 1992, *Proposed Spokane resource management plan amendment and final environmental impact statement:* U.S. Bureau of Land Management, Report no. 920286, 181 p.

U.S. Bureau of Land Management, 1994, *Fort Wainwright, Yukon maneuver area, proposed resource management plan and final environmental impact statement,* Fairbanks North Star Borough, Alaska: U.S. Bureau of Land Management, Report no. 940001, 124 p.

U.S. Bureau of Land Management, 1994, *resource management plan and environmental impact statement for the Grass Creek resource area,* Worland District, Wyoming: U.S. Bureau of Land Management, Technical Report, 299 p.

U.S. Bureau of Land Management, 1994, *Stateline resource management plan, Clark and Nye Counties,* Nevada (draft environmental impact statement and Draft Supplement): U.S. Bureau of Land Management, Report no. 940177.

U.S. Bureau of Land Management, 1996, *resource management plan for the Green River resource area, Rock Springs,* Wyoming: U.S. Bureau of Land Management, Report no. 960148.

U.S. Bureau of Land Management, 1997, *Proposed resource management plan/final environmental impact statement for the Roswell resource area,* Roswell, New Mexico and proposed resource management plan Amendment/Final environmental impact statement for the Carlsbad resource area, Carlsbad, New Mexico. Volume 1: U.S. Bureau of Land Management, Report no. BLMNMPT970031610V1, 379 p.

U.S. Bureau of Land Management, 1997, *Proposed resource management plan/final environmental impact statement for the Roswell resource area,* Roswell, New Mexico and proposed resource management plan Amendment/Final environmental impact statement for the Carlsbad resource area, Carlsbad, New Mexico. Volume 2: U.S. Bureau of Land Management, Report no. BLMNMPT970031610V2, 535 p.

U.S. Bureau of Land Management, 1998, *Draft Caliente management framework plan amendment and environmental impact statement for management of desert tortoise habitat,* Lincoln County, Nevada: U.S. Bureau of Land Management, Report no. 980190, 322 p.

U.S. Bureau of Land Management, 1998, *Judith-Valley-Phillips resource management plan—* Chouteau, Fergus, Judith Basin, Petroleum, Phillips, and Valley Counties, Montana (draft supplement to the final environmental impact statement of September 1997): U.S. Bureau of Land Management, Report no. 980149, 51 p.

U.S. Bureau of Land Management, 1999, *Final environmental impact statement for management of desert tortoise habitat,* Lincoln County, Nevada: U.S. Bureau of Land Management, Report no. 990201, 442 p.

U.S. Bureau of Land Management, 2000, *Federal fluid minerals leasing and development, Otero and Sierra Counties,* New Mexico: U.S. Bureau of Land Management, Report no. 000383, 521 p.

U.S. Bureau of Land Management, 2000, *Pinedale anticline oil and gas exploration and development project,* Sublette County, Wyoming: U.S. Bureau of Land Management, Report no. 000159, 301 p.

U.S. Bureau of Land Management, 2001, *National management strategy for motorized off-highway vehicle use on public lands:* U.S. Bureau of Land Management, Report no. PB2001103 162, 58 p.

U.S. Bureau of Land Management, 2003, *Farmington resource management plan,* San Juan, McKinley, Rio Arriba, and Sandoval Counties, New Mexico: U.S. Bureau of Land Management, Report no. 030143.

U.S. Bureau of Land Management, 2004, *Dillon resource management plan, Beaverhead and Madison Counties,* Montana: U.S. Bureau of Land Management, Report no. 040153, 626 p.

U.S. Bureau of Land Management, 2004, *Jack Morrow Hills coordinated activity plan, Sweetwater, Fremont, and Sublette Counties,* Wyoming: U.S. Bureau of Land Management, Technical Report.

U.S. Bureau of Land Management, 2005, *Newmont Mining Company, emigrant project, Elko County,* Nevada: U.S. Bureau of Land Management, Report no. 050117, 269 p.

U.S. Bureau of Land Management, 2006, *Coeur d'Alene resource management plan, Idaho:* U.S. Bureau of Land Management, Report no. 060002, 347 p.

U.S. Bureau of Reclamation, 2001, *Pothole Reservoir resource management plan, Grant County,* Washington: U.S. Bureau of Reclamation, Report no. 010518, 567 p.

U.S. Department of Defense, 1999, *Military training in the Marianas, Guam and the Commonwealth of the Northern Marianas:* U.S. Department of Defense, Report no. 990187, 478 p.

U.S. Department of the Navy, 1973, *Off-road vehicle use of Mirror Lake Naval Weapons Center,* China Lake, California: U.S. Department of the Navy, Report no. ELR73–141 1, 67 p.

U.S. Federal Highway Administration, 2005, *US 33 Nelsonville bypass, city of Nelsonville, Hocking and Athens Counties,* Ohio: U.S. Federal Highway Administration, Report no. 050282, 828 p.

U.S. Fish and Wildlife Service, 1985, *Alaska Peninsula National Wildlife Refuge final comprehensive conservation plan,* environmental impact statement, and wilderness review: U.S. Fish and Wildlife Service, Technical Report, 426 p.

U.S. Fish and Wildlife Service, 1986, *Great Dismal Swamp National Wildlife Refuge Master Plan,* Virginia and North Carolina: U.S. Fish and Wildlife Service, Report no. 860501, 245 p.

U.S. Forest Service, 1973, *Off-road vehicle policy, Hoosier National Forest, Indiana: U.S. Forest Service,* Report no. USDAFS–DES(ADM)–73–51, 110 p.

U.S. Forest Service, 1973, *Proposed off-road vehicle regulations and administrative instructions:* U.S. Forest Service, Report no. USDAFS–FES(ADM)–73–49, 110 p.

U.S. Forest Service, 1973, *Proposed regulations and administrative instructions relating to use of off-road vehicles on National Forest lands:* U.S. Forest Service, Report no. USDAFS–DES(ADM)–73–49, 22 p.

U.S. Forest Service, 1973, *Unit plan for management of the South Holston unit: U.S. Forest Service,* Report no. USDAFS–DES(ADM)–73–50, 80 p.

U.S. Forest Service, 1985, *Caribou National Forest and Curlew National Grassland land and resource management plan,* Idaho, Utah, and Wyoming: U.S. Forest Service, Report no. 850437.

U.S. Forest Service, 1985, *Kisatchie National Forest land and resource management plan,* Louisiana: U.S. Forest Service, Report no. 850414.

U.S. Forest Service, 1985, *Proposed land and resource management plan for the Medicine Bow National Forest and Thunder Basin National Grassland,* Wyoming: U.S. Forest Service, Report no. 850523.

U.S. Forest Service, 1986, *Chequamegon National Forest, Wisconsin land and resource management plan:* U.S. Forest Service, Report no. 860325.

U.S. Forest Service, 1986, *Cleveland National Forest land and resource management plan,* Orange, Riverside, and San Diego Counties, California: U.S. Forest Service, Report no. 860209.

U.S. Forest Service, 1986, *Croatan and Uwharrie National Forests land and resource management plan,* North Carolina: U.S. Forest Service, Report no. 860218.

U.S. Forest Service, 1986, *Proposed Land and resource management plan for Wayne National Forest,* Ohio: U.S. Forest Service, Report no. 860426.

U.S. Forest Service, 1987, *Angeles National Forest land and resource management plan,* Los Angeles, Ventura, and San Bernardino Counties, California: U.S. Forest Service, Report no. 870394.

U.S. Forest Service, 1987, *Proposed Apache-Sitgreaves National Forests Land and resource management plan,* Apache, Coconino, Greenlee, and Navajo Counties, Arizona: U.S. Forest Service, Report no. 870387.

U.S. Forest Service, 1988, *Proposed Kaibab National Forest land and resource management plan,* Coconino, Yavapai, and Mohave Counties, Arizona: U.S. Forest Service, Report no. 880109.

U.S. Forest Service, 1989, *San Bernardino National Forest land and resource management plan,* San Bernardino and Riverside Counties, California: U.S. Forest Service, Report no. 890023.

U.S. Forest Service, 1991, *Corral Mountain timber sale, San Juan National Forest,* Archuleta County, Colorado: U.S. Forest Service, Report no. 910124, 59 p.

U.S. Forest Service, 1992, *Norbeck Wildlife Preserve land management plan,* Black Hills National Forest, Custer and Pennington Counties, South Dakota (final supplement to the environmental impact statement of July 1989): U.S. Forest Service, Report no. 920352, 92 p.

U.S. Forest Service, 1992, *Rangeland ecosystem management in the Uinta National Forest,* Provo, Utah: U.S. Forest Service, Report no. 920322, 189 p.

U.S. Forest Service, 1992, *Sierra National Forest land and resource management plan,* Fresno, Madera, and Mariposa Counties, California: U.S. Forest Service, Report no. 920105.

U.S. Forest Service, 1993, *Six Rivers National Forest plan,* Humboldt, Del Norte, Siskiyou, and Trinity Counties, California: U.S. Forest Service, Report no. 930340.

U.S. Forest Service, 1994, *Oregon Dunes National Recreation Area management plan,* Siuslaw National Forest—Coos, Douglas, and Lane Counties, Oregon: U.S. Forest Service, Report no. 940292.

U.S. Forest Service, 1994, *Revised land and resource management plan for the national forests and grasslands in Texas*—Angelina, Fannin, Houston, Jasper, Montague, Montgomery, Nacogdoches, Newton, Sabine, San Augustine, San Jacinto, Shelby, Trinity, Walker, and Wise Counties, Texas: U.S. Forest Service, Report no. 940385.

U.S. Forest Service, 1995, *Draft environmental impact statement, Fish Bate analysis area*—North Fork Ranger District, Clearwater National Forest, Clearwater County, Idaho: U.S. Forest Service, Technical Report.

U.S. Forest Service, 1995, *East Fork Blacks Fork analysis area, Wasatch-Cache National Forest, Summit County,* Utah (final supplement to the final environmental impact statement of July 1992): U.S. Forest Service, Report no. 950294, 49 p.

U.S. Forest Service, 1995, *Mendocino National Forest land and resource management plan*—Colusa, Glenn, Lake, Mendocino, Tehama, and Trinity Counties, California: U.S. Forest Service, Report no. 950337.

U.S. Forest Service, 1995, *Snowy Trail re-route, Ventura County,* California: U.S. Forest Service, Report no. 950117, 109 p.

U.S. Forest Service, 1996, *Mendenhall Glacier Recreation Area management plan Revision, Chatham area,* Juneau Ranger District, Tongass National Forest, Alaska: U.S. Forest Service, Report no. 960205, 279 p.

U.S. Forest Service, 1996, *Proposed Revised Land and resource management plan for Francis Marion National Forest, Berkeley and Charleston Counties,* South Carolina: U.S. Forest Service, Report no. 960158.

U.S. Forest Service, 1996, *Snowy Trail re-route, Los Padres National Forest, Ventura County, California:* U.S. Forest Service, Report no. 960524.

U.S. Forest Service, 2001, *Land and resource management plan, Uinta National Forest—Juab,* Sanpete, Toole, Utah, and Wasatch Counties, Utah: U.S. Forest Service, Report no. 010142.

U.S. Forest Service, 2001, *Starbucky restoration project, Red River Ranger District, Nez Perce National Forest,* Idaho County, Idaho: U.S. Forest Service, Report no. 010274.

U.S. Forest Service, 2002, *Final environmental impact statement amendment to the Land and resource management plan*—Management direction for acquired lands in southeastern Oklahoma, Ouachita National Forest, McCurtain County, Oklahoma: U.S. Forest Service, Technical Report.

U.S. Forest Service, 2002, *Uncompahgre National Forest travel plan revision,* Gunnison, Hinsdale, Mesa, Montrose, Ouray, San Juan, and San Miguel Counties, Colorado (final supplement to the final environmental impact statement of April 2000): U.S. Forest Service, Report no. 020131, 92 p.

U.S. Forest Service, 2002, *West gold project, Sandpoint Ranger District,* Idaho Panhandle National Forests, Bonner County, Idaho: U.S. Forest Service, Report no. 020484.

U.S. Forest Service, 2002, *Whiskey Campo resource management project,* Mountain Home Ranger District, Boise National Forest, Elmore County, Idaho: U.S. Forest Service, Report no. 020299, 151 p.

U.S. Forest Service, 2003, *Big Bend Ridge vegetation management project and timber Sale,* Ashton/Island Park Ranger District, Caribou-Targhee National Forest, Freemont County, Idaho: U.S. Forest Service, Report no. 030225.

U.S. Forest Service, 2003, *Duck Creek–Swains access management project,* Cedar City Ranger District, Dixie National Forest, Iron, Garfield, and Kane Counties, Utah: U.S. Forest Service, Report no. 030343, 49 p.

U.S. Forest Service, 2003, *Upper Bear timber sale, Council Ranger District, Payette National Forest,* Adams County, Idaho: U.S. Forest Service, Report no. 030319, 478 p.

U.S. Forest Service, 2004, *Duck Creek fuels treatment analysis, Cedar City Ranger District, Dixie National Forest,* Kane County, Utah: U.S. Forest Service, Report no. 040535, 240 p.

U.S. Forest Service, 2004, French Face ecosystem restoration, Ninemile Ranger District, Lolo National Forest, Montana: U.S. Forest Service, Report no. 040336, 184 p.

U.S. Forest Service, 2005, Dean Project Area, Bearlodge Ranger District, Black Hills National Forest, Crook County, Wyoming: U.S. Forest Service, Report no. 050212, 192 p.

U.S. Forest Service, 2005, Final environmental impact statement for the access designation on the Ocala National Forest, Lake, Marion, and Putnam Counties, Florida: U.S. Forest Service, Technical Report.

U.S. Forest Service, 2005, Gallatin National Forest travel management plan, Gallatin, Madison, Park, Meagher, Sweetgrass, and Carbon Counties, Montana: U.S. Forest Service, Report no. 050231.

U.S. Forest Service, 2005, Rocky Mountain Ranger District travel management plan, Rocky Mountain Ranger District, Lewis and Clark National Forest, Glacier, Pondera, Teton, and Lewis and Clark Counties, Montana: U.S. Forest Service, Report no. 050236, 386 p.

U.S. Forest Service, 2005, Woodrock project, Tongue Ranger District, Bighorn National Forest, Sheridan County, Wyoming: U.S. Forest Service, Report no. 050112, 201 p.

U.S. Forest Service, 2006, Hoosier National Forest land and resource management plan—Final environmental impact statement: U.S. Forest Service, Technical Report.

U.S. Immigration and Naturalization Service, 2004, U.S. Border Patrol activities within the border areas of the Tuscon and Yuma Sectors, Arizona (Revision of the Draft environmental impact statement of November 2002): U.S. Immigration and Naturalization Service, Report no. 040479, 487 p.

U. S. National Park Service, 1979, Proceedings of the conference on scientific research in the National Parks, San Francisco, California, November 26–30, 1979, U.S. Forest Service, Washington D.C.

U.S. National Park Service, 1991, Big Cypress National Preserve, Collier, Monroe, and Dade Counties, Florida: U.S. National Park Service, Report no. 910386.

U.S. National Park Service, 1998, Redwood National and State Parks general management plan, Del Norte and Humboldt Counties, California: U.S. National Park Service, Report no. 980290, 474 p.

U.S. National Park Service, 1999, Big Cypress National Preserve, recreational off-road vehicle management plan—Collier, Dade, and Monroe Counties, Florida (draft supplement to the environmental impact statement of October 1991): U.S. National Park Service, Report no. 990285, 171 p.

U.S. National Park Service, 2000, Recreational Off-road Vehicle management plan, Big Cypress National Preserve—Collier, Miami-Dade, and Monroe Counties, Florida (final supplement to the final environmental impact statement of October 1991): U.S. National Park Service, Report no. 000275, 619 p.

Valentin, C., Poesen, J., and Li, Y., 2005, *Gully erosion—Impacts, factors and control: Catena,* v. 63, no. 2–3, p. 132–153.

Walker, D.A., and Everett, K.R., 1987, Road dust and its environmental impact on Alaskan taiga and tundra: *Arctic and Alpine Research,* v. 19, no. 4, p. 479–489.

Walker, D.A., Webber, P.J., Everett, K.R., and Brown, J., 1977, The effects of low-pressure wheeled vehicles on plant communities and soils at Prudhoe Bay, Alaska: Hanover, New Hampshire, *Cold Regions Research and Engineering Lab,* Report no. CRREL–SR–77–17, 50 p.

Watson, J.J., Kerley, G.I.H., and McLachlan, A., 1996, *Human activity and potential impacts on dune breeding birds in the Alexandria Coastal Dunefield:* Landscape and Urban Planning, v. 34, no. 3–4, p. 3 15–322.

Webb, R.H., Wilshire, H.G., and Henry, M.A., 1983, *Natural recovery of soils and vegetation following human disturbance, Environmental effects of off-road vehicles—Impacts and management in arid regions*: New York, Springer-Verlag, p. 279–302.

Welch, R., Madden, M., and Doren, R.F., 1999, Mapping the Everglades: *Photogrammetric Engineering and Remote Sensing,* v. 65, no. 2, p. 163–170.

Wester, L., 1994, *Weed management and the habitat protection of rare species—A case study of the endemic Hawaiian fern Marsilea villosa: Biological Conservation,* v. 68, no. 1, p. 1–9.

Whitten, K.R., and Cameron, R.D., 1983, Movements of collared caribou, *Rangifer tarandus,* in relation to petroleum development on the Arctic slope of Alaska: *Canadian Field-Naturalist,* v. 97, no. 2, p. 143–146.

Wiedemann, A.M., 1984, *Ecology of Pacific Northwest coastal sand dunes—A community profile: Evergreen State College,* Olympia, Washington, Report no. FWSOBS8404, 146 p.

Wilcox, D.A., 1989, Migration and control of purple loosestrife (*Lythrum salicaria* L.) along highway corridors: *Environmental Management,* v. 13, no. 3, p. 365–370.

Wilshire, H.G., 1983, Off-road vehicle recreation management policy for public lands in the United States—A case history: *Environmental Management,* v. 7, no. 6, p. 489–499.

Wilshire, H.G., and Nakata, J.K., 1976, *Off-road vehicle effects on California's Mojave Desert: California Geology,* June 1976, p. 123–132.

Wilshire, H.G., Nakata, J.K., Shipley, Susan, and Prestegaard, Karen, 1978, Impacts of vehicles on natural terrain at seven sites in the San Francisco Bay area: *Environmental Geology,* v. 2, no. 5, p. 295–3 19.

Wilshire, H.G., Shipley, Susan, and Nakata, J.K., 1978, Impacts of off-road vehicles on vegetation: *Transactions of the North American Wildlife and Natural Resources Conference,* v. 43, p. 131–139.

Winner, W.E., and Atkinson, C.J., 1986, Absorption of air pollutants by plants and consequences: *Trends in Ecology and Evolution,* v. 1, p. 15–18.

1.3. OHV Effects on Wildlife and Habitats: Native, Threatened, and Endangered Species

Adams, E.S., 1975, Effects of lead and hydrocarbons from snowmobile exhaust on brook trout (*Salvelinus fontanlis*): *Transactions of the American Fisheries Society,* v. 104, no. 2, p. 363–373.

Adams, J.A., Stolzy, L.H., Endo, A.S., Rowlands, P.G., and Johnson, H.B., 1982, Desert soil compaction reduces annual plant cover: *California Agriculture,* v. 36, no. 9–10, p. 6–7.

Adams, L.W., and Dove, L.E., 1989, Wildlife reserves and corridors in the urban environment— *A guide to ecological landscape planning and resource conservation:* Columbia, Maryland, National Institute for Urban Wildlife, 91 p.

Adams, L.W., and Geis, A.D., 1983, Effects of roads on small mammals: *Journal of Applied Ecology,* v. 20, no. 2, p. 403–415.

Alaska Pulp Corporation, U.S. Forest Service, and U.S. *Army Corps of Engineers,* 1994, Ushk Bay timber sale, Alaska Pulp Corporation long-term timber sale Contract, Chatnam Area, Tongass National Forest, Alaska: U.S. Forest Service and U.S. Army Corps of Engineers, Report no. 940408.

Albrecht, J., and Knopp, T.B., 1985, Off road vehicles—Environmental impact—Management response—*A bibliography: Rosemount, Minnesota,* Agricultural Experiment Station, University of Minnesota, Miscellaneous Publication, 50 p.

Algers, B., Ekesbo, I., and Strömberg, S., 1978, *The impact of continuous noise on animal health: Acta Veterinaria Scandinavica (Supplementum),* v. 67, p. 1–26.

Andersen, D.E., Rongstad, O.J., and Mytton, W.R., 1986, The behavioral response of red-tailed hawk to military training activity: *Journal of Raptor Research,* v. 20, p. 65–68.

Andrews, A., 1990, Fragmentation of habitat by roads and utility corridors—A review: *Australian Zoologist,* v. 26, p. 130–141.

Ashley, E.P., and Robinson, J.T., 1996, Road mortality of amphibians, reptiles, and other wildlife on the Long Point Causeway, Lake Erie, Ontario: *The Canadian Field-Naturalist,* v. 110, p. 403–412.

Assistant Secretary of Defense (Health and Environment), 1973, Proposed Department of Defense regulation on use of off-road vehicles: *U.S. Department of Defense,* Report no. ELR0511, 20 p.

Awbrey, F.T., Hunsaker, D., and Church, R., 1995, Acoustical responses of California gnatcatchers to traffic noise: *Inter-noise,* v. 65, p. 97 1–974.

Bakowski, C., and Kozakiewicz, M., 1988, The effect of forest road on bank vole and yellow-necked mouse populations: *Acta Theriologica,* v. 33, no. 12–25, p. 345–353.

Baur, A., and Baur, B., 1990, Are roads barriers to dispersal in the land snail *Arianta arbustorum*?: *Canadian Journal of Zoology,* v. 68, no. 33, p. 613–617.

Becking, R.W., and Hayes, G.L., 1970, *Public land policy and the environment*—Volume 3, part II (cont'd)—Environmental problems on the public lands—Case studies 9 through 17: Denver, Colorado, Rocky Mountain Center on Environment, Report no. PLLRC29–3, 436 p.

Beier, P., 1993, Determining minimum habitat areas and habitat corridors for cougars: *Conservation Biology,* v. 7, no. 11, p. 94–108.

Beier, P., 1996, *Metapopulation models, tenacious tracking, and cougar conservation, in* McCullough, D.R., ed., Metapopulations and wildlife conservation: Washington, D.C., Island Press, p. 293–323.

Belnap, Jayne, 1995, Surface disturbances—Their role in accelerating desertification: *Environmental Monitoring and Assessment,* v. 37, no. 1–3, p. 39–57.

Belnap, Jayne, 2003, The world at your feet—Desert biological soil crusts: *Frontiers in Ecology and the Environment,* v. 1, no. 4, p. 181–189.

Ben Brooks and Associates, Ironhorse Investors, Inc., Santa Fe Pacific Railroad Company, and U.S. Bureau of Land Management, 1998, *Hualapai Mountain Land Exchange,* Mohave County, Arizona: Report no. 980482, 175 p.

Berry, K.H., 1980, A review of the effects of off-road vehicles on birds and other vertebrates, *in* DeGraaf, R.M., and Tilghman, N.G., eds., Management of western forests and grasslands for nongame birds—Workshop proceedings, Salt Lake City, Utah, February 11–14, 1980: Ogden Utah, U.S. Forest Service, *Intermountain Forest and Range Experiment Station,* General Technical Report INT–86, p. 451–467.

Berry, K.H., 1980, The effects of four-wheel vehicles on biological resources, *in* Andrews, R.N.L., and Nowak, P., eds., Off-road vehicle use: a management challenge: Washington, D.C., U.S. *Office of Environmental Quality,* p. 231-233.

Berry, K.H., Shields, T., Woodman, A.P., Campbell, T., Roberson, J., Bohuski, K., and Karl, A., 1986, Changes in desert tortoise populations at the Desert Tortoise Research Natural Area between 1979 and 1985, Desert Tortoise Council proceedings of symposium 10 1986, Palmdale, California, March 22-24, 1986: *Desert Tortoise Council,* p. 100–123.

Benson, P.E., Nokes, W.A., Cramer, R.L., O'Connor, J., Lindeman, W., 1986, Air-quality and noise issues in environmental planning: Washington, D.C., Transportation Research Board and National Research Council Technical Report, 72 p., *http://www.osti.gov/energycitations/product.biblio.jsp?osti_id=7172765.*

Bjornlie, D.D., and Garrott, R.A., 2001, Effects of winter road grooming on bison in Yellowstone National Park: *Journal of Wildlife Management,* v. 65, no. 3, p. 560–572.

Boarman, W.I., Beigel, M.L., Goodlett, G.C., and Sazaki, M., 1998, A passive integrated transponder system for tracking animal movements: *Wildlife Society Bulletin,* v. 26, no. 4, p. 886–891.

Boarman, W.I., and Sazaki, M., 1996, Highway mortality in desert tortoises and small vertebrates—Success of barrier fences and culverts, *in* Evink, G.L., Garrett, P., Zeigler, D., and Berry, J., eds., Trends in addressing transportation related wildlife mortality—Proceedings of the transportation related wildlife mortality seminar, Orlando, Florida, April 30–May 2, 1996: Tallahassee, Florida, State Department of Transportation, Report no. FL-ER-58-96, *http://www. icoet. net/ICO WET/96proceedings. asp.*

Boarman, W.I., and Sazaki, M., 2006, A highway's road-effect zone for desert tortoises (*Gopherus agassizii*): *Journal of Arid Environments,* v. 65, no. 1, p. 94–101.

Boarman, W.I., Sazaki, M., Berry, K.H., Goodlett, G.O., Jennings, W.B., and Woodman, A.P., 1992, Measuring the effectiveness of a tortoise-proof fence and culverts—Status report from first field season, *Proceedings of the Desert Tortoise Council: Desert Tortoise Council,* p. 126–142.

Boarman, W.I., Sazaki, M., and Jennings, W.B., 1997, The effects of roads, barrier fences, and culverts on desert tortoise populations in California, USA, *in* Abbeema, J.V., ed., *Proceedings—Conservation, restoration, and management of tortoises and turtles—An international conference, State University of New York, Purchase*, New York, July 1993: Orange, New Jersey, New York Turtle and Tortoise Society, p. 54–58.

Boer, B., 1998, Anthropogenic factors and their potential impacts on the sustainable development of Abu Dhabi's terrestrial biological resources: *International Journal of Sustainable Development and World Ecology,* v. 5, no. 2, p. 125–135.

Bolling, J.D., and Walker, L.R., 2000, Plant and soil recovery along a series of abandoned desert roads: *Journal of Arid Environments,* v. 46, no. 11, p. 1–24.

Bond, J., 1971, Noise—Its effect on the physiology and behavior of animals: *Agricultural Science Review*, v. 9, no. 4, p. 1–10.

Bowles, A.E., 1995, Responses of wildlife to noise, *in* Knight, R.L., and Gutzwiller, K.J., eds., *Wildlife and recreationists—Coexistence through management and research:* Washington, D.C., Island Press, p. 109–156.

Boyle, S.A., and Samson, F.B., 1985, Effects of nonconsumptive recreation on wildlife—A review: *Wildlife Society Bulletin,* v. 13, p. 110–116.

Braby, R.J., Shapira, A., and Simmons, R.E., 2001, Successful conservation measures and new breeding records for Damara Terns Sterna balaenarum in Namibia: *Marine Ornithology,* v. 29, no. 2, p. 8 1–84.

Brattstrom, B.H., 2000, The range, habitat requirements, and abundance of the orange-throated whiptail, *Cnemidophorus hyperythrus beldingi*: *Bulletin of the Southern California Academy of Sciences,* v. 99, no. 1, p. 1–24.

Brattstrom, B.H., and Bondello, M.C., 1983, Effects of off-road vehicle noise on desert vertebrates, *in* Webb, R.H., and Wilshire, H.G., eds., *Environmental effects of off-road vehicles—Impacts and management in arid regions:* New York, Springer-Verlag, p. 167–206.

Brezonick, M., 1996, *Designing for the sound of silence: Diesel Progress Engines and Drives,* v. 62, no. 8, p. 4.

Brillinger, D.R., Preisler, H.K., Ager, A.A., and Wisdom, M.J., 2004, Stochastic differential equations in the analysis of wildlife motion, *2004 Proceedings of the American Statistical*

Association, section on statistics and the environment, Toronto, Ontario, Canada, August 8–12, 2004: Alexandria, Virginia, American Statistical Association.

Brody, A.J., and Pelton, M.R., 1989, Effects of roads on black bear movements in western North Carolina: *Wildlife Society Bulletin,* v. 17, p. 5–10.

Brooks, M.L., 1995, Benefits of protective fencing to plant and rodent communities of the western Mojave Desert, California: *Environmental Management,* v. 19, no. 1, p. 65–74.

Brooks, M.L., 1999, Effects of protective fencing on birds, lizards, and black-tailed hares in the western Mojave Desert: *Environmental Management,* v. 23, no. 3, p. 387–400.

Brown, G., and Porembski, S., 2000, Phytogenic hillocks and blow-outs as 'safe sites' for plants in an oil-contaminated area of northern Kuwait: *Environmental Conservation,* v. 27, no. 3, p. 242–249.

Bruinderink, G.W.T.A., and Hazebroek, E., 1986, Ungulate traffic collisions in Europe: *Conservation Biology,* v. 10, no. 4, p. 1059–1067.

Buckley, R., and King, N., 2003, Visitor-impact data in a land-management context, *in* Buckley, R.C., Pickering, D., and Weaver, D.B., eds., *Nature-based tourism, environment and land management:* Wallingford, England, CABI Publishing, p. 89-99.

Buick, A.M., and Paton, D.C., 1989, *Impact of off-road vehicles on the nesting success of hooded plovers Charadrius rubricollis in the Coorong region of South Australia:* Emu, v. 89, no. 3, p. 159–172.

Bury, R.B., 1980, What we know and do not know about off-road vehicle impacts on wildlife, *in* Andrews, R.N.L., and Nowak, P., eds., Off-road vehicle use—A management challenge: Washington, D.C., U.S. *Office of Environmental Quality,* p. 110-122.

Bury, R.B., and Luckenbach, R.A., 2002, Comparison of desert tortoise (*Gopherus agassizii*) populations in an unused and off-road vehicle area in the Mojave Desert: *Chelonian Conservation and Biology,* v. 4, no. 2, p. 457–463.

Bury, R.B., Luckenbach, R.A., and Busack, S.D., 1977, Effects of off-road vehicles on vertebrates in the California desert USA: Washington, D.C., U.S. Fish and Wildlife Service, Report, *Wildlife Research Reports* No. 8, p. 1–23.

Busack, S.D., and Bury, R.B., 1974, Some effects of off-road vehicles and sheep grazing on lizard populations in the Mojave Desert: Biological Conservation, v. 6, no. 3, p. 179–183.

Butcher, J., Majer, J.D., and Unsworth, P., 1989, Bibliography of fauna studies in reclaimed lands, *in* Majer, J.D., ed., *Animals in primary succession—The role of fauna in reclaimed lands:* New York, Cambridge University Press, p. 451-516.

Cimon, N.J., and Wisdom, M.J., 2004, Accurate velocity estimates from inaccurate GPS data, Remote sensing for field users—*Proceedings of the tenth Forest Service remote sensing applications conference,* Salt Lake City, Utah, April 5–9, 2004.

Clark, W.D., and Karr, J.R., 1979, Effects of highways on red-winged blackbird and horned lark populations: *The Wilson Bulletin,* v. 91, p. 143–145.

Collins, E., O'Farrell, T.P., and Rhoads, W., 1982, *Annotated bibliography for biologic overview for the Nevada nuclear waste storage investigations,* Nevada Test Site, Nye County, Nevada: Goleta, California, EG and G, Inc., Report no. EGG1 1832419, 48 p.

Collins, E., O'Farrell, T.P., and Rhoads, W., 1982, *Biologic overview for the Nevada Nuclear Waste Storage investigations,* Nevada Test Site, Nye County, Nevada: Goleta, California, EG and G, Inc., Report no. EGG1 1832460, 55 p.

Dahle, J.L., 1983, *Research study on the technology for control of environmental noise from off- road haul trucks*: San Diego, California, Woodward Associates, Inc., Report no. BUMINESOFR–100–84, 94 p.

Delaney, D.K., Grubb, T.G., Beiber, P., Pater, L.L., and Reiser, M.H., 1999, Effects of helicopter noise on Mexican spotted owls: *Journal of Wildlife Management*, v. 63, no. 1, p. 60–76.

Delforce, R.J., Sinden, J.A., and Young, M.D., 1986, Policy preferences and social economic values to resolve pastoralism—*Tourism conflicts: Landscape Planning,* v. 12, no. 4, p. 387–401.

Dennis, D.F., 1998, Analyzing public inputs to multiple objective decisions on national forests using conjoint analysis: *Forest Science,* v. 44, no. 3, p. 421–428.

Dhindsa, M.S., Sandhu, J.S., Sandhu, P.S., and Toor, H.S., 1988, Roadside birds in Punjab (India)—Relation to mortality from vehicles: *Environmental Conservation,* v. 15, no. 4, p. 303–310.

Doak, D., Kareiva, P., and Klepetka, B., 1994, Modeling population viability for the desert tortoise in the western Mojave Desert: *Ecological Applications,* v. 4, no. 3, p. 446–460.

Dobson, A., Ralls, K., Foster, M., Soulé, M.E., Simberloff, D., Doak, D., Estes, J.A., Mills, L.S., Mattson, D., Dizro, R., Arita, H., Ryan, S., Norse, E.A., Noss, R.F., and Johns, D., 1999, Connectivity—Maintaining flows in fragmented landscapes, *in* Soulé, M.E., and Terborgh, J., eds., *Continental Conservation—Scientific foundations of regional reserve networks:* Washington, D.C., Island Press, p. 129–170.

Dunn, J.P., Summerfield, C.J., and Johnson, M., 2003, Distribution, seasonal cycle, host plant records, and habitat evaluation of a Michigan threatened insect—The Great Plains spittlebug, *Lepyronia gibbosa* (Homoptera: Cercopidae): *Great Lakes Entomologist,* v. 35, no. 2, p. 121–129.

Ehrenfeld, J.G., and Schneider, J.P., 1991, *Chamaecyparis thyoides* wetlands and suburbanization—Effects on hydrology, water quality and plant community composition: *Journal of Applied Ecology,* v. 28, no. 2, p. 467–490.

Evink, G.L., Garrett, P., Zeigler, D., and Berry, J., 1996, Trends in addressing transportation related wildlife mortality—*Proceedings of the transportation related wildlife mortality seminar,* Orlando, Florida, April 30–May 2, 1996: Tallahassee, Florida, State Department of Transportation, Report no. FL-ER-58-96, *http://www.icoet.net/ICOWET/ 96proceedings. asp.*

Evink, G.L., Garrett, P., Zeigler, D., and Berry, J., eds., 1998, Proceedings of the International Conference on Wildlife Ecology and Transportation, Fort Myers, Florida, February 10–12, 1998: Tallahassee, Florida, Florida Department of Transportation, *Environmental Management Office Report no.* FL DOT FL-ER 69-98.

Fatoki, O.S., 2000, Trace zinc and copper concentrations in roadside vegetation and surface soils—A measurement of local atmospheric pollution in Alice, South Africa: *International Journal of Environmental Studies* (UK), v. 57, no. 5, p. 501–5 13.

Feldhammer, G.A., Gates, J.E., Harman, D.M., Loranger, A.J., and Dixon, K.R., 1986, Effects of interstate highway fencing on white-tailed deer activity: *Journal of Wildlife Management,* v. 50, no. 497–503.

Felix, N.A., Raynolds, M.K., Jorgenson, J.C., and Dubois, K.E., 1992, Resistance and resilience of tundra plant communities to disturbance by winter seismic vehicles: *Arctic and Alpine Research,* v. 24, no. 1, p. 69–77.

Forman, R.T.T., and Alexander, L.E., 1998, Roads and their major ecological effects: *Annual Review of Ecology and Systematics,* v. 29, p. 207–231.

Foster, M.L., and Humphrey, S.R., 1995, Use of highway underpasses by Florida panthers and other wildlife: *Wildlife Society Bulletin,* v. 23, no. 1, p. 95–100.

Furniss, M.J., Flanagan, S.A., and McFadin, B.A., 2000, Hydrologically connected roads—An indicator of the influence of roads on chronic sedimentation, surface water hydrology, and exposure to toxic chemicals: *U.S. Forest Service, Stream Systems Technology Center,* Rocky Mountain Research Station, Technical Report, 4 p.

Garland, T., and Bradley, W.G., 1984, Effects of a highway on Mojave Desert rodent populations: *American Midland Naturalist,* v. 111, p. 47–56.

Gibbs, J.P., 1998, Amphibian movements in response to forest edges, roads, and streambeds in southern New England: *Journal of Wildlife Management,* v. 62, no. 2, p. 584–589.

Gibeau, M.L., and Herrero, S., 1998, Roads, rails, and grizzly bears in the Bow River Valley, Alberta, *in* Evink, G.L., Garrett, P., Zeigler, D., and Berry, J., eds., *Proceedings of the International Conference on Wildlife Ecology and Transportation,* Fort Myers, Florida, February 10–12, 1998: Tallahassee, Florida, Florida Department of Transportation, Environmental Management Office Report no. FL DOT FL-ER 69-98, p. 104–108.

Gish, C.D., and Christensen, R.E., 1973, Cadmium, nickel, lead, and zinc in earthworms from roadside soil: *Environmental Science and Technology,* v. 7, p. 1060–1062.

Glass, R.L., 1986, Hydrologic conditions in Connors Bog area, Anchorage, Alaska: U.S. Geological Survey, *Technical Report,* 23 p.

Godfrey, P.J., and Godfrey, M.M., 1981, Ecological effects of off-road vehicles on Cape Cod: *Oceanus*, v. 23, no. 4, p. 56–57.

Grant, T.J., 2005, Flat-tailed horned lizards (*Phrynosoma mcallii*)—Population size estimation, effects of off-highway vehicles, and natural history: Fort Collins, Colorado, *M.S. thesis,* Department of Wildlife Biology, Colorado State University, 84 p., *http://www.warnercnr.colostate.edu/~doherty/Tyler%20Grant%20M.S.%20Thesis_FINA L.p df*

Greene, A.F.C., 1975, The need for cooperative approaches to fish and wildlife management: *Transactions of the North American Wildlife and Natural Resources Conference*, v. 40, p. 133–141.

Grue, C.E., O'Shea, T.J., and Hoffman, D.J., 1984, Lead concentrations and reproduction in highway-nesting barn swallows [*Hirundo rustica*]: *Condor,* v. 86, no. 4, p. 383–389.

Gutzwiller, K.J., and Barrow, W.C., 2003, Influences of roads and development on bird communities in protected Chihuahuan desert landscapes: *Biological Conservation,* v. 113, no. 2, p. 225–237.

Hall, C., and Dearden, P., eds., 1984, The impact of "non-consumptive" recreation on wildlife— An annotated bibliography: Monticello, Illinois, *Vance Bibliographies,* 45 p.

Hansen, R.P., Hillhouse, W.A., Willard, B.E., and Burke, H.D., 1970, Public land policy and the environment. Volume 2. Part II—Environmental problems on the public lands. *Summary statement and case studies 1 through 8:* Denver, Colorado, Rocky Mountain Center on Environment, Report no. PLLRC29–2, 409 p.

Hanski, I., 1999, *Metapopulation ecology:* New York, Oxford University Press, 324 p.

Hanski, I., and Simberloff, D., 1997, The metapopulation approach, its history, conceptual domain, and application to conservation, *in* Hanski, L., Gilpin, M., and Hanski, I., eds.,

Metapopulation biology—Ecology, genetics, and evolution: San Diego, California, Academic Press, p. 5–26.

Harris, L.D., and Gallagher, P.B., 1989, New initiatives for wildlife conservation—The need for movement corridors, *in* Mackintosh, G., ed., *Preserving communities and corridors:* Washington, D.C., Defenders of Wildlife, p. 11–34.

Hastings, M.C., 1995, *Physical effects of noise on fishes, Inter-noise 95, the 1995 International Congress on Noise Control,* July 10, 1995, p. 979–984.

Haupt, H.F., 1959, Road and slope characteristics affecting sediment movement from logging roads: *Journal of Forestry,* v. 57, no. 5, p. 329–339.

Hayes, D.L., 1991, The all-American canal lining project—A catalyst for rational and comprehensive groundwater management on the United States-Mexico border: *Natural Resources Journal,* v. 31, no. 4, p. 803–827.

Hendriks, R.W., 1989, *Traffic noise attenuation as a function of ground vegetation: California Department of Transportation,* Report no. FHWA/CA/TL–89/09, 99 p.

Hockey, P.A.R., 1983, The distribution, population size, movements and conservation of the African black oystercatcher *Haematopus moquini*: *Biological Conservation, v.* 25, no. 3, p. 233–262.

Holsman, R.H., 2005, Management opportunities and obligations for mitigating off-road vehicle impacts to wildlife and their habitats: *Transactions of the North American Wildlife and Natural Resources Conference,* v. 70, p. 399–417.

Holzapfel, C., and Schmidt, W., 1990, Roadside vegetation along transects in the Judean Desert: *Israel Journal of Botany,* v. 39, p. 263–270.

Hornby, N., and Sheate, W.R., 2001, Sustainable management of recreational off-road vehicles in National Parks in the UK: *Environmental and Waste Management,* v. 4, no. 2, p. 95–106.

Hosier, P.E., Kochlar, M., and Thayer, V., 1981, Off-road vehicle and pedestrian track effects on the sea-approach of hatchling loggerhead turtles: *Environmental Conservation,* v. 8, no. 2, p. 158–161.

Huey, L.M., 1941, Mammalian invasion via the highway: *Journal of Mammalogy,* v. 22, p. 383–385.

Hunt, A., Dickens, H.J., and Whelan, R.J., 1987, Movement of mammals through tunnels under railway lines: *Australian Zoologist,* v. 24, no. 2, p. 89–93.

Hyers, A.D., and Marcus, M.G., 1981, Land use and desert dust hazards in central Arizona, *in* Pewe, T., ed., Desert dust—Origin, characteristics, and effect on man: Boulder, Colorado, *Geological Society of America,* Inc., p. 267–280.

Irwin, C.L., Garrett, P., and McDermott, K.P., 2003 *Proceedings of the International Conference on Ecology And Transportation,* Lake Placid, New York, August 24–29, 2003: Raleigh, North Carolina, Center for Transportation and the Environment, North Carolina State University.

Irwin, C.L., Garrett, P., and McDermott, K.P., 2005, *Proceedings of the 2005 International Conference On Ecology And Transportation,* San Diego, California, August 29–September 2, 1995: Raleigh, North Carolina, Center for Transportation and the Environment, North Carolina State University.

Jensen, O., 1997, *Adaptive strategies in a mountain valley—A case study from Baltistan: Copenhagen,* Institute of Geography, University of Copenhagen, 88 p.

Johnson, R.R., and Simpson, J.M., 1988, *Desertification of wet riparian ecosystems in arid regions of the North American southwest, in* Whitehead, E.E., Hutchinson, C.F.,

Timmermann, B.N., and Varady, R.G., eds., *Arid lands—Today and tomorrow, Tuscon, Arizona,* October 20–25, 1985: Boulder, Colorado, Westview Press, p. 1383–1393.

Karanja, G., 2003, *Tourism impacts in Masai Mara National Reserve:* IIED Wildlife and Development Series, v. 14, p. 5–16.

Kaseloo, P.A., and Tyson, K.O., 2004, Synthesis of noise effects on wildlife populations: U.S. Federal Highway Administration, *Office of Research and Technology Services,* Report no. FHWA–HEP–06–016, 67 p.

Kato, T.T., and O'Farrell, T.P., 1986, *Biological assessment of the effects of petroleum production at maximum efficient rate,* Naval Petroleum Reserve No. 1 (Elk Hills), Kern County, California, on the endangered blunt-nosed leopard lizard, *Gambelia silus*: Goleta, California, EG and G Energy Measurements, Inc., Report no. EGG102 822 108, 72 p.

Katz, M., Legore, R.S., Weitkamp, D., Cummins, J.M., and Anderson, D., 1972, Effects on freshwater fish: *Journal of the Water Pollution Control Federation,* v. 44, no. 6, p. 1226–1250.

Kay, B.L., and Graves, W.L., 1983, Revegetation and stabilization techniques for disturbed desert vegetation, *in* Webb, R.H., and Wilshire, H.G., eds., *Environmental effects of off-road vehicles—Impacts and management in arid regions*: New York, Springer-Verlag, p. 3 25–340.

Kellert, S.R., 1980, *Activities of the American public relating to animals.* Phase II: U.S. Department of Interior Fish and Wildlife Service, Technical Report.

Khalaf, F.I., 1989, Desertification and aeolian processes in the Kuwait Desert: Journal of Arid Environments, v. 16, no. 2, p. 125–145.

Kline, N.C., and Swann, D.E., 1998, Quantifying wildlife road mortality in Saguaro National Park, *in* Evink, G.L., Garrett, P., Zeigler, D., and Berry, J., eds., *Proceedings of the International Conference on Wildlife Ecology and Transportation,* Fort Myers, Florida, February 10–12, 1998: Tallahassee, Florida, Florida Department of Transportation, Environmental Management Office Report no. FL DOT FL-ER 69-98, p. 23–31.

Knight, R.L., and Gutzwiller, K.J., 1995, *Wildlife and recreationists—Coexistence through management and research:* Washington, D.C, Island Press, 372 p.

Knight, R.L., and Kawashima, J.Y., 1993, Responses of raven and red-tailed hawk populations to linear right-of-ways: *Journal of Wildlife Management,* v. 57, no. 2, p. 265–271.

Kurczerski, F.E., 2000, *History of white pine (Pinus strobus)/oak (Quercus spp.) savanna in southern Ontario, with particular reference to the biogeography and status of the antenna- waving wasp, Tachysphex pechumani* (Hymenoptera: Sphecidae): Canadian Field-Naturalist, v. 114, no. 1, p. 1–20.

Kurczewski, F.E., 1998, Distribution, status, evaluation, and recommendations for the protection of *Tachysphex pechumani* Krombein, the antennal-waving wasp: *Natural Areas Journal,* v. 18, no. 3, p. 242–254.

Lacey, R.M., Goettel, R.G., Balbach, H.E., and Severinghaus, W.D., 1982, *Planning for off-road recreational vehicle use on Army installations: Champaign, Illinois, Construction Engineering Research Lab,* Technical Report, 98 p.

Lacey, R.M., and Severinghaus, W.D., 1982, *Evaluation of lands for off-road recreational four- wheel drive vehicle use: U.S. Department of Commerce Government Reports Announcements and Index National Technical Information Service (NTIS),* Technical Report, 1755 p.

Land, D., and Lotz, M., 1996, Wildlife crossing designs and use by Florida panthers and other wildlife in southwest Florida, *in* Evink, G.L., Garrett, P., Zeigler, D., and Berry, J., eds., Trends in addressing transportation related wildlife mortality—*Proceedings of the transportation related wildlife mortality seminar,* Orlando, Florida, April 30–May 2, 1996: Tallahassee, Florida, State Department of Transportation, Report no. FL-ER-58-96, *http://www. icoet. net/ICO WET/96proceedings. asp.*

Langdon, A.M., 2000, Mojave Desert soils, plants, and ants—*Developing a monitoring strategy for off-highway-vehicles:* Pomona, California, California State Polytechnic University, Pomona, 114 p.

Lehnert, M.E., and Bissonette, J.A., 1997, Effectiveness of highway crosswalk structures at reducing deer-vehicle collisions: *Wildlife Society Bulletin,* v. 25, no. 4, p. 809–818.

Lightfoot, D.C., and Whitford, W.G., 1991, Productivity of creosotebush foliage and associated canopy arthropods along a desert roadside: *American Midland Naturalist,* v. 125, no. 2, p. 310–322.

Lodico, N.J., 1973, *Environmental effects of off-road vehicles—A review of the literature: U.S. Office of Library Services,* Research Services Branch, Technical Report, 109 p.

Lopez, J.J., and West, N., 1994, *Management issues concerning the piping plover and off-road vehicles at Cape Cod National Seashore, Coastal Zone Canada '94—Cooperation in the coastal zone,* Halifax, Nova Scotia, Canada, September 1994: Halifax, Nova Scotia, Canada, Ocean Studies, Office of Research Services, A&A, and Dalhousie University.

Lovallo, M.J., and Anderson, E.M., 1996, Bobcat (*Lynx rufus*) home range size and habitat use in northwest Wisconsin: *American Midland Naturalist,* v. 135, no. 2, p. 241–252.

Lovallo, M.J., and Anderson, E.M., 1996, Bobcat movements and home ranges relative to roads in Wisconsin: *Wildlife Society Bulletin,* v. 24, no. 1, p. 7 1–76.

Lovich, J.E., and Bainbridge, D., 1999, Anthropogenic degradation of the southern California desert ecosystem and prospects for natural recovery and restoration: *Environmental Management,* v. 24, no. 3, p. 309–326.

Luckenbach, R.A., 1978, An analysis of off-road vehicle use on desert avifaunas: *Transactions of the North American Wildlife and Natural Resources Conference,* v. 43, p. 157–162.

Luckenbach, R.A., 1982, *Ecology and management of the desert tortoise (Gopherus agassizii) in California: U.S. Fish and Wildlife Service,* Technical Report, 37 p.

Luckenbach, R.A., and Bury, R.B., 1983, Effects of off-road vehicles on the biota of the Algodones Dunes, Imperial County, California, USA: *Journal of Applied Ecology,* v. 20, no. 1, p. 265–286.

Lyren, L., 2001, *Movement patterns of coyotes and bobcats relative to roads and underpasses in the Chino Hills area of southern California:* Pomona, California, California State Polytechnic University, Pomona, 127 p.

Mader, H.J., 1984, *Animal habitat isolation by roads and agricultural fields: Biological Conservation,* v. 29, no. 1, p. 8 1–96.

McDonald, P.M., and Plumb, G.E., 1996, *Black-footed ferret reintroduction in the Conata Basin/Badlands of southwestern South Dakota: Rangelands,* v. 18, no. 6, p. 222–224.

Meffe, G.K., and Carroll, R.C., 1997, Principles of conservation biology: Sunderland, Massachusetts, *Sinauer Associates,* Inc., 729 p.

Melvin, S.M., Griffin, C.R., and MacIvor, L.H., 1991, Recovery strategies for piping plovers in managed coastal landscapes: *Coastal Management,* v. 19, no. 1, p. 2 1–34.

Melvin, S.M., Hecht, A., and Griffin, C.R., 1994, Piping plover mortalities caused by off-road vehicles on Atlantic coast beaches: *Wildlife Society Bulletin,* v. 22, no. 3, p. 409–414.

Memphis State University, 1971, *Effects of noise on wildlife and other animals: U.S. Environmental Protection Agency,* Office of Noise Abatement and Control, Report no. NTID300.5, 77 p.

Mize, R., Evans, R.E., MacRoberts, B.R., MacRoberts, M.H., and Rudolph, D.C., 2005, Restoration of pitcher plant bogs in eastern Texas, USA: *Natural Areas Journal,* v. 25, no. 2, p. 197–201.

Morgan, M.T., 1993, Nutrition is the key to plight of an ancient survivor, the desert tortoise: *Smithsonian Institution Research Reports,* v. 74, p. 4.

Mumme, R.L., Schoech, S.J., Woolfenden, G.W., and Fitzpatrick, J.W., 2000, Life and death in the fast lane—Demographic consequences of road mortality in the Florida scrub-jay: *Conservation Biology,* v. 14, no. 2, p. 501–5 12.

Munguira, M.L., and Thomas, J.A., 1992, Use of road verges by butterfly and burnet populations, and the effect of roads on adult dispersal and mortality: *Journal of Applied Ecology,* v. 29, p. 316–329.

Naylor, L.M., 2006, *Behavioral responses of Rocky Mountain elk (Cervus elaphus) to recreational disturbance:* Corvallis, OR, Oregon State University, 103 p.

Noor, M., and Shah, B.H., 1995, Infiltration capacity and soil bulk density differences between grazed area and off-road track at Paya (Kaghan Valley): *Pakistan Journal of Forestry,* v. 45, no. 1, p. 13–18.

Noss, R.F., and Cooperrider, A.Y., 1994, *Saving nature's legacy—Protecting and restoring biodiversity:* Washington, D.C., Island Press, 417 p.

Noss, R.F., and Csuti, B., 1997, Habitat fragmentation, *in* Meffe, G.K., and Carroll, C.R., eds., *Principles of conservation biology:* Sunderland, Massachusetts, Sinauer Associates, Inc., p. 269–304.

O'Farrell, T.P., 1984, Conservation of the endangered San Joaquin kit fox *Vulpes macrotis mutica* in the Naval Petroleum Reserves, California: *Acta Zoologica Fennica,* v. 172, p. 207–208.

O'Farrell, T.P., and Gilbertson, L., 1986, Ecology of the desert kit fox *Vulpes macrotis arsipus* in the Mojave Desert of southern California USA: *Bulletin of the Southern California Academy of Sciences,* v. 85, no. 1, p. 1–15.

O'Farrell, T.P., Harris, C.E., Kato, T.T., and McCue, P.M., 1986, *Biological assessment of the effects of petroleum production at maximum efficient rate, Naval Petroleum Reserve No.* 1 (Elk Hills), Kern County, California, on the endangered San Joaquin kit fox, *Vulpes macrotis mutica*: Goleta, California, EG and G Energy Measurements, Inc., Report no. EGG102822107, 84 p.

O'Farrell, T.P., and Kato, T.T., 1987, *Biological assessment of the effects of petroleum production activities, Naval Petroleum Reserves in California, on the endangered giant kangaroo rat, Dipodomys ingens: Goleta, California, EG and G Energy Measurements,* Inc., Report no. EGG102822183, 34 p.

Oxley, D.J., Fenton, M.B., and Carmody, G.R., 1974, Effects of roads on populations of small mammals: *Journal of Applied Ecology,* v. 11, no. 1, p. 51–59.

Paquet, P., and Callahan, C., 1996, Effects of linear developments of winter movement of gray wolves in the Bow River Valley of Banff National Park, Alberta, *in* Evink, G.L., Garrett, P., Zeigler, D., and Berry, J., eds., Trends in addressing transportation related wildlife mortality——*Proceedings of the transportation related wildlife mortality seminar,* Orlando, Florida, April 30–May 2, 1996: Tallahassee, Florida, State Department of Transportation, Report no. FL-ER-58-96, *http://www.icoet.net/ICOWET/96proceedings. asp.*

Parendes, L.A., and Jones, J.A., 2000, Role of light availability and dispersal in exotic plant invasion along roads and streams in the H. J. Andrews Experimental Forest, Oregon: *Conservation Biology,* v. 14, no. 1, p. 64–75.

Piehl, B.T., Beschta, R.L., and Pyles, M.R., 1988, Ditch-relief culverts and low-volume forest roads in the Oregon Coast Range: *Northwest Science,* v. 62, no. 3, p. 9 1–98.

Power, J.A., and Schuster, E.G., 1975, *Off-road vehicle law and Idaho—An ORV planning aid:* Moscow, Idaho, University of Idaho, Technical Report, 65 p.

Preisler, H.K., Ager, A.A., and Wisdom, M.J., 2006, Statistical methods for analysing responses of wildlife to human disturbance: *Journal of Applied Ecology,* v. 43, no. 1, p. 164–172.

Propst, D.B., Shomaker, J.H., and Mitchekkm, J.E., 1977, *Attitudes of Idaho off-road vehicle users and mangers:* Moscow, Idaho, College of Forestry, Wildlife and Range Sciences, University of Idaho, Technical Report, 30 p.

Quarles, H.D., Hanawalt, R.B., and Odum, W.E., 1974, Lead in small mammals, plants, and soil at varying distances from a highway: *Journal of Applied Ecology,* v. 11, no. 3, p. 937–949.

Rafique, S.A., 1994, Effect of grazing management and fertilizer application on vegetation and soil properties of a moist temperate forest range in Siran Valley (Mansehra), NWFP: *Pakistan Journal of Forestry,* v. 44, no. 1, p. 20–29.

Reed, R.A., Johnson-Barnard, J., and Baker, W.L., 1996, Contribution of roads to forest fragmentation in the Rocky Mountains: *Conservation Biology,* v. 10, no. 4, p. 1098–1106.

Reijnen, R., and Foppen, R., 1995, The effects of car traffic on breeding bird populations in woodland. IV. Influence of population size on the reduction of density close to a highway: *Journal of Applied Ecology,* v. 32, no. 3, p. 481–491.

Reijnen, R., Foppen, R., Braak, C.T., and Thissen, J., 1995, The effects of car traffic on breeding bird populations in woodland. III. Reduction of density in relation to the proximity of main roads: *Journal of Applied Ecology,* v. 32, no. 1, p. 187–202.

Reijnen, R., Foppen, R., and Veenbaas, G., 1997, Disturbance by traffic of breeding birds—Evaluation of the effect and considerations in planning and managing road corridors: *Biodiversity and Conservation,* v. 6, no. 4, p. 567–581.

Rhoads, W.A., Cochrane, S.A., and Williams, M.P., 1979, *Status of endangered and threatened plant species on Tonopah Test Range—A survey:* Goleta, California, Santa Barbara Research Center, Report no. SAND807035, 126 p.

Rickard, W.H., 1988, *Natural vegetation at the proposed reference repository location in southeastern Washington:* Richland, Washington, Battelle Pacific Northwest Labs, Report no. PNL–6402, 27 p.

Rinne, J.N., Rickel, B., and Hendrickson, D., 1980, A new Gila topminnow locality in southern Arizona: Fort Collins, Colorado, *U.S. Forest Service, Rocky Mountain Forest and Range Experiment Station,* Report no. RM-3 82, 4 p.

Romin, L.A., and Bissonette, J.A., 1996, *Temporal and spatial distribution of highway mortality of mule deer on newly constructed roads at Jordanelle Reservoir,* Utah: Great Basin Naturalist, v. 56, no. 1, p. 1–11.

Roni, P., Beechie, T.J., Bilby, R.E., Leonetti, F.E., Pollock, M.M., and Pess, G.R., 2002, A review of stream restoration techniques and a hierarchical strategy for prioritizing restoration in Pacific Northwest watersheds: *North American Journal of Fisheries Management,* v. 22, no. 1, p. 1–20.

Rosen, P.C., and Lowe, C.H., 1994, Highway mortality of snakes in the Sonoran Desert of southern Arizona: *Biological Conservation,* v. 68, no. 22, p. 143–148.

Rowland, M.M., Wisdom, M.J., Johnson, B.K., and Penninger, M.A., 2005, Effects of roads on elk—Implications for management in forested ecosystems, *in* Wisdom, M.J., ed., *The Starkey Project—A synthesis of long-term studies of elk and mule deer:* Lawrence, Kansas, Alliance Communications Group, p. 42–52.

Roy, A.H., Rosemond, A.D., Leigh, D.S., Paul, M.J., and Wallace, J.B., 2003, Habitat-specific responses of stream insects to land cover disturbance—Biological consequences and monitoring implications: *Journal of the North American Benthological Society,* v. 22, no. 2, p. 292–307.

Rudolph, D.C., Burgdorf, S.J., Conner, R.N., and Dickson, J.G., 1998, The impact of roads on the timber rattlesnake, (*Crotalus horridus*), in eastern Texas, *in* Evink, G.L., Garrett, P., Zeigler, D., and Berry, J., eds., *Proceedings of the International Conference on Wildlife Ecology and Transportation,* Fort Myers, Florida, February 10–12, 1998: Tallahassee, Florida, Florida Department of Transportation, Environmental Management Office Report no. FL DOT FL-ER 69-98, p. 236–240.

Ruediger, B., 1998, Rare carnivores and highways—Moving into the 21st century, *in* Evink, G.L., Garrett, P., Zeigler, D., and Berry, J., eds., *Proceedings of the International Conference on Wildlife Ecology and Transportation,* Fort Myers, Florida, February 10–12, 1998:

Tallahassee, Florida, Florida Department of Transportation, *Environmental Management Office Report* no. FL DOT FL-ER 69-98, p. 10–16.

Samways, M.J., 1989, Insect conservation and landscape ecology—*A case-history of bush crickets (Tettigoniidae) in southern France: Environmental Conservation,* v. 16, no. 3, p. 217–226.

Saunders, D.A., Hobbs, R.J., and Margules, C.R., 1991, Biological consequences of ecosystem fragmentation—A review: *Conservation Biology,* v. 5, no. 1, p. 18–32.

Schipani, S.P., Bruno, R.S., Lattin, M.A., King, B.M., and Patton, D.J., 1998, *Quantification of cognitive process degradation while mobile, attributable to the environmental stressors endurance, vibration, and noise:* Aberdeen Proving Ground, Maryland, U.S. Army Research Lab, Report no. ARL–TR–1 603, 96 p.

Seibert, H.C., and Conover, J.H., 1991, Mortality of vertebrates and invertebrates on an Athens County, Ohio [USA], highway: *Ohio Journal of Science,* v. 91, no. 4, p. 163–166.

Servheen, C., Walker, J., and Kasworm, W., 1998, Fragmentation effects of high-speed highways on grizzly bear populations shared between the United States and Canada, *in* Evink, G.L., Garrett, P., Zeigler, D., and Berry, J., eds., *Proceedings of the International Conference on Wildlife Ecology and Transportation,* Fort Myers, Florida, February 10–12, 1998: Tallahassee, Florida, Florida Department of Transportation, Environmental Management Office Report no. FL DOT FL-ER 69-98, p. 97–103.

Sheridan, D., 1979, Off-road vehicles on public land: Washington, D.C., *Council on Environmental Quality,* Report no. PB8621 1158, 94 p.

Singer, F.J., 1978, Behavior of mountain goats in relation to US Highway 2, Glacier Park, Montana: *Journal of Wildlife Management,* v. 42, p. 59 1–597.

Smith, W.H., 1976, Lead contamination of the roadside ecosystem: *Journal of the Air Pollution Control Association,* v. 26, p. 753–766.

Spellerburg, I.F., 1998, Ecological effects of roads and traffic—*A literature review: Global Ecology and Biogeography Letters,* v. 7, no. 5, p. 3 17–333.

Spellerburg, I.F., and Morrison, T., 1998, The ecological effects of new roads—*A literature review:* Wellington, New Zealand, New Zealand Department of Conservation, Technical Report, 55 p.

Spencer, H.J., and Port, G.R., 1988, Effects of roadside conditions on plants and insects: II. Soil conditions: *Journal of Applied Ecology,* v. 25, no. 22, p. 709–7 15.

Steiner, A.J., and Leatherman, S.P., 1979, *An annotated bibliography of the effects of off-road vehicle and pedestrian traffic on coastal ecosystems: U.S. National Park Service Cooperative Research unit and University of Massachusetts Amherst,* Report no. UMNPSCRU45, 88 p.

Steiner, A.J., and Leatherman, S.P., 1981, Recreational impacts on the distribution of ghost crabs *Ocypode quadrata* Fab: *Biological Conservation,* v. 20, no. 2, p. 111–122.

Stroh, T., 2001, *Proposed Collbran rock and soil source project—Wildlife report: U.S. Bureau of Reclamation,* Report no. PB2006101846, 28 p.

Sullivan, B.K., 2000, Long-term shifts in snake populations—A California site revisited: *Biological Conservation,* v. 94, no. 3, p. 321–325.

Swihart, R.K., and Slade, N.A., 1984, Road crossing in *Sigmodon hispidus* and *Microtus ochrogaster*: *Journal of Mammalogy,* v. 65, no. 2, p. 357–360.

Tewes, M.E., and Blanton, D.R., 1998, Potential impacts of international bridges on ocelots and jaguarundis along the Rio Grande wildlife corridor, *in* Evink, G.L., Garrett, P., Zeigler, D., and Berry, J., eds., *Proceedings of the International Conference on Wildlife Ecology and Transportation,* Fort Myers, Florida, February 10–12, 1998: Tallahassee, Florida, Florida Department of Transportation, Environmental Management Office Report no. FL DOT FL-ER 69-98, p. 135–139.

Trombulak, S.C., and Frissell, C.A., 2000, Review of ecological effects of roads on terrestrial and aquatic communities: *Conservation Biology,* v. 14, no. 1, p. 18–30.

Tshiguvho, T.E., Dean, W.R.J., and Robertson, H.G., 1999, Conservation value of road verges in the semi-arid Karoo, South Africa—Ants (Hymenoptera: Formicidae) as bio-indicators: *Biodiversity and Conservation,* v. 8, no. 12, p. 1683–1695.

Twiss, R., Sidener, J., Bingham, G., Burke, J.E., and Hall, C.H., 1980, *Potential impacts of geothermal development on outdoor recreational use of the Salton Sea:* Livermore, California, University of California, Livermore, Technical Report, 61 p.

Tyser, R.W., and Worley, C.A., 1992, Alien flora in grasslands adjacent to road and trail corridors in Glacier National Park, Montana (U.S.A.). *Conservation Biology,* v. 6, no. 2, p. 253-262.

U.S. Army Corps of Engineers, 1994, Stationing of mechanized or armored combat forces at Fort Lewis, Thurston and Pierce Counties, Washington: U.S. *Army Corps of Engineers, Report no. 940045.*

U.S. Army Corps of Engineers, 1998, *McGregor range land withdrawal, Fort Bliss, Otera County,* New Mexico: U.S. Army Corps of Engineers, Report no. 980439, 586 p.

U.S. Army Corps of Engineers, 2001, *Fort Bliss mission and master plan—Dona Ana and Otero Counties,* New Mexico, and El Paso County, Texas: U.S. Army Corps of Engineers, Report no. 010081.

U.S. Army Corps of Engineers, and Oklahoma Department of Wildlife Conservation, 1986, Candy Lake, Candy Creek, Osage County, Oklahoma (supplemental information report to the final environmental impact statement of January 1975): *U.S. Army Corps of Engineers,* Report no. 860246, 4 p.

U.S. Bureau of Land Management, 1984, Book Cliffs resource management plan, Utah: *U.S. Bureau of Land Management,* Report no. 840541, 534 p.

U.S. Bureau of Land Management, 1984, Proposed Monument resource management plan, Idaho: *U.S. Bureau of Land Management,* Report no. 840584, 419 p.

U.S. Bureau of Land Management, 1985, Box Elder resource management plan, Utah: *U.S. Bureau of Land Management,* Report no. 850448.

U.S. Bureau of Land Management, 1985, Grand Junction resource area—resource management plan and environmental impact statement: *U.S. Bureau of Land Management,* Report, 171 p.

U.S. Bureau of Land Management, 1985, Lemhi resource management plan, Idaho: *U.S. Bureau of Land Management,* Report no. 850426.

U.S. Bureau of Land Management, 1985, Lower Gila South resource management plan, La Paz, Maricopa, Pima, Pinal, and Yuma Counties, Arizona: *U.S. Bureau of Land Management,* Report no. 850347, 909 p.

U.S. Bureau of Land Management, 1985, Medicine Lodge Proposed resource management plan, Bingham, Bonneville, Butte, Clark, Fremont, Jefferson, Madison, and Teton Counties, Idaho: *U.S. Bureau of Land Management,* Report no. 850237.

U.S. Bureau of Land Management, 1985, Proposed Two Rivers resource management plan, Oregon: *U.S. Bureau of Land Management,* Report no. 850409.

U.S. Bureau of Land Management, 1985, resource management plan for the Kemmerer resource area, Lincoln, Sublette, Sweetwater, and Uinta Counties, Wyoming: *U.S. Bureau of Land Management,* Report no. 850532.

U.S. Bureau of Land Management, 1985, resource management plan for the Walker Planning Area, Nevada: *U.S. Bureau of Land Management,* Report no. 850274.

U.S. Bureau of Land Management, 1985, Rio Puerco resource management plan, New Mexico: *U.S. Bureau of Land Management,* Report no. 850417.

U.S. Bureau of Land Management, 1985, Two Rivers resource management plan, Oregon: *U.S. Bureau of Land Management,* Report no. 850137, 163 p.

U.S. Bureau of Land Management, 1985, White Sands resource area resource management plan, New Mexico: *U.S. Bureau of Land Management,* Report no. 850387.

U.S. Bureau of Land Management, 1985, Yuma District resource management plan, Yuma, La Paz, and Mohave Counties, Arizona, and San Bernardino, Riverside, and Imperial Counties, California: *U.S. Bureau of Land Management,* Report no. 850358, 310 p.

U.S. Bureau of Land Management, 1986, Baker resource management plan, Baker County, Oregon: *U.S. Bureau of Land Management,* Report no. 860150, 129 p.

U.S. Bureau of Land Management, 1986, Big Lost/Pahsimeroi Wilderness, Idaho: *U.S. Bureau of Land Management,* Report no. 860399, 125 p.

U.S. Bureau of Land Management, 1986, Eastern San Diego County planning unit, California— Preliminary wilderness recommendations: *U.S. Bureau of Land Management*, Report no. 860388, 195 p.

U.S. Bureau of Land Management, 1986, Lander resource management plan, Lander, Wyoming: *U.S. Bureau of Land Management,* Report no. 860449.

U.S. Bureau of Land Management, 1986, Little Snake resource management plan, Moffat, Rio Blanco, and Routt Counties, Colorado: *U.S. Bureau of Land Management,* Report no. 860029.

U.S. Bureau of Land Management, 1986, Proposed 1985 amendments to the California desert conservation area plan and the Eastern San Diego County Master Framework Plan, California: *U.S. Bureau of Land Management,* Report no. 860432, 138 p.

U.S. Bureau of Land Management, 1986, Proposed 1985 amendments to the California desert conservation area plan, California: *U.S. Bureau of Land Management,* Report no. 860077.

U.S. Bureau of Land Management, 1986, Proposed Baker resource management plan, Vale District, Oregon: *U.S. Bureau of Land Management,* Report no. 860396.

U.S. Bureau of Land Management, 1986, resource management plan for the Carlsbad resource area, Eddy, Lea, and Chaves Counties, New Mexico: *U.S. Bureau of Land Management,* Report no. 860101, 390 p.

U.S. Bureau of Land Management, 1986, resource management plan for the House Range resource area, Juab and Millard Counties, Utah: *U.S. Bureau of Land Management,* Report no. 860377.

U.S. Bureau of Land Management, 1986, resource management plan for the Warm Springs resource area, Millard County, Utah: *U.S. Bureau of Land Management,* Report no. 860395.

U.S. Bureau of Land Management, 1986, Utah statewide wilderness environmental impact statement: *U.S. Bureau of Land Management,* Report no. 860036.

U.S. Bureau of Land Management, 1986, Wilderness Suitability EIS—Grass Creek and Cody resource areas, Bighorn Basin, Wyoming (draft supplement for the Owl Creek Wilderness study area): *U.S. Bureau of Land Management,* Report no. 860168, 40 p.

U.S. Bureau of Land Management, 1987, Canon City District wilderness study, Colorado: *U.S. Bureau of Land Management,* Report no. 870444, 238 p.

U.S. Bureau of Land Management, 1987, *Cascade resource management plan,* Idaho: U.S. Bureau of Land Management, Report no. 870281, 432 p.

U.S. Bureau of Land Management, 1987, Glenwood Springs resource area resource management plan, Colorado—Wilderness recommendations: U.S. *Bureau of Land Management,* Report no. 870421, 168 p.

U.S. Bureau of Land Management, 1987, Medicine Bow-Divide resource areas resource management plan, Rawlins District, Wyoming: *U.S. Bureau of Land Management,* Report no. 870212.

U.S. Bureau of Land Management, 1987, North Dakota resource management plan, Dunn and Bowman Counties, North Dakota: *U.S. Bureau of Land Management,* Report no. 870236, 199 p.

U.S. Bureau of Land Management, 1987, Pinedale resource area resource management plan, Sublette and Lincoln Counties, Wyoming: *U.S. Bureau of Land Management*, Report no. 870077, 347 p.

U.S. Bureau of Land Management, 1987, Pinedale resource area resource management plan, Sublette, Teton, Lincoln, and Fremont Counties, Wyoming: *U.S. Bureau of Land Management,* Report no. 870428, 232 p.

U.S. Bureau of Land Management, 1987, Pocatello resource management plan, Idaho: *U.S. Bureau of Land Management,* Report no. 870333, 132 p.

U.S. Bureau of Land Management, 1987, Preliminary wilderness recommendations for the Arcata resource area, Eden Valley and Thatcher Ridge Wilderness Study Areas, Ukiah District, Mendocino County, California: *U.S. Bureau of Land Management,* Report no. 870348, 111 p.

U.S. Bureau of Land Management, 1987, Proposed Lemhi resource management plan, Lemhi County, Idaho—Eighteenmile Wilderness: *U.S. Bureau of Land Management,* Report no. 870437, 178 p.

U.S. Bureau of Land Management, 1987, Proposed Medicine Lodge, Idaho resource management plan—Wilderness recommendations: *U.S. Bureau of Land Management,* Report no. 870436, 92 p.

U.S. Bureau of Land Management, 1987, Proposed Monument resource management plan and Wilderness Study, Idaho: *U.S. Bureau of Land Management,* Report no. 870248, 162 p.

U.S. Bureau of Land Management, 1987, Proposed resource management plan and wilderness recommendations for the Lahontan resource area, Nevada: *U.S. Bureau of Land Management,* Report no. 870293, 282 p.

U.S. Bureau of Land Management, 1987, Proposed resource management plan for the San Juan resource area, Moab District, Utah: *U.S. Bureau of Land Management,* Report no. 870440.

U.S. Bureau of Land Management, 1987, Proposed Utility Corridor resource management plan, Alaska: *U.S. Bureau of Land Management,* Report no. 870291, 281 p.

U.S. Bureau of Land Management, 1987, Proposed Wilderness Program for the Stafford District Wilderness EIS Area, Cochise, Gila, Graham, and Greenlee Counties, Arizona and Hidalgo County, New Mexico: *U.S. Bureau of Land Management,* Report no. 870156, 502 p.

U.S. Bureau of Land Management, 1987, Proposed wilderness program for the Upper Sonoran Wilderness EIS Area, Maricopa, Mohave, Yavapai, and Yuma Counties, Arizona: *U.S. Bureau of Land Management,* Report no. 870292, 394 p.

U.S. Bureau of Land Management, 1987, resource management plan for the Elko resource area, Nevada—Wilderness Plan: *U.S. Bureau of Land Management,* Report no. 870400, 172 p.

U.S. Bureau of Land Management, 1987, Taos resource management plan, New Mexico: *U.S. Bureau of Land Management,* Report no. 870331.

U.S. Bureau of Land Management, 1987, Washakie resource area resource management plan, portions of Big Horn, Hot Springs, and Washakie Counties, Wyoming: *U.S. Bureau of Land Management,* Report no. 870395.

U.S. Bureau of Land Management, 1987, Wilderness recommendations for the north-central California Wilderness Study Areas: *U.S. Bureau of Land Management,* Report no. 870058, 115 p.

U.S. Bureau of Land Management, 1987, Winnemucca District wilderness recommendations, Nevada and California: *U.S. Bureau of Land Management,* Report no. 870342, 581 p.

U.S. Bureau of Land Management, 1988, Brothers/Lapine resource management plan, Oregon: *U.S. Bureau of Land Management,* Report no. 880365.

U.S. Bureau of Land Management, 1988, Central California Section 202 wilderness study areas wilderness recommendations—Sheep Ridge, Milk Ranch/Case Mountain, and Ventana contiguous wilderness study areas: *U.S. Bureau of Land Management,* Report no. 880020, 110 p.

U.S. Bureau of Land Management, 1988, Challis resource area proposed resource management plan, upper Columbia-Salmon Clearwater Districts, Custer and Lemhi Counties, Idaho: *U.S. Bureau of Land Management,* Report no. 980508.

U.S. Bureau of Land Management, 1988, Cody resource area resource management plan, Big Horn and Park Counties, Wyoming: *U.S. Bureau of Land Management,* Report no. 880315.

U.S. Bureau of Land Management, 1988, Fort Greely National maneuver area and Fort Greely Air Drop Zone, Alaska—Resource management plan: *U.S. Bureau of Land Management,* Report no. 880276, 138 p.

U.S. Bureau of Land Management, 1988, Fort Wainwright maneuver area, Fairbanks North Star Borough, Alaska: *U.S. Bureau of Land Management,* Report no. 880277, 138 p.

U.S. Bureau of Land Management, 1988, Medicine Bow-Divide resource areas (now Great Divide resource area) resource management plan, Rawlins District, Wyoming: *U.S. Bureau of Land Management,* Report no. 880325.

U.S. Bureau of Land Management, 1988, Pony Express resource management plan, Toole, Utah, and Salt Lake Counties, Utah: *U.S. Bureau of Land Management,* Report no. 880310, 147 p.

U.S. Bureau of Land Management, 1988, Proposed Las Vegas resource management plan and final environmental impact statement, Clark and Nye Counties, Nevada: *U.S. Bureau of Land Management,* Report no. 980207.

U.S. Bureau of Land Management, 1988, Uncompahgre Basin resource management plan, Colorado: *U.S. Bureau of Land Management,* Report no. 880345.

U.S. Bureau of Land Management, 1988, West Hiline resource management plan, northcentral Montana: *U.S. Bureau of Land Management,* Report no. 880200, 306 p.

U.S. Bureau of Land Management, 1988, Wilderness recommendations for the Red Mountain Wilderness Study Area in the Arcata resource area, California: *U.S. Bureau of Land Management*, Report no. 880033, 120 p.

U.S. Bureau of Land Management, 1988, Wilderness Study Areas in the Rock Springs District, Fremont, Lincoln, Sublette, and Sweetwater Counties, Wyoming (revised draft environmental impact statement): *U.S. Bureau of Land Management*, Report no. 880359, 175 p.

U.S. Bureau of Land Management, 1989, Arcata resource management plan, California: *U.S. Bureau of Land Management*, Report no. 890306, 354 p.

U.S. Bureau of Land Management, 1989, Dixie resource area management plan, Washington County, Utah: *U.S. Bureau of Land Management*, Report no. 890290, 250 p.

U.S. Bureau of Land Management, 1989, Phoenix resource management plan, Arizona: *U.S. Bureau of Land Management*, Report no. 890012, 240 p.

U.S. Bureau of Land Management, 1989, Powder River wilderness suitability study for the Powder River resource area resource management plan, Miles City District, Montana: *U.S. Bureau of Land Management*, Report no. 890301, 73 p.

U.S. Bureau of Land Management, 1989, San Pedro River riparian management plan for the San Pedro River EIS Area, Cochise County, Arizona: *U.S. Bureau of Land Management*, Report no. 890152, 381 p.

U.S. Bureau of Land Management, 1989, San Rafael resource management plan, Emery County, Utah: *U.S. Bureau of Land Management*, Report no. 890240.

U.S. Bureau of Land Management, 1989, Utility Corridor Proposed resource management plan, Alaska: *U.S. Bureau of Land Management*, Report no. 890326, 281 p.

U.S. Bureau of Land Management, 1989, White Sands resource management plan amendment for McGregor Range, Otero County, New Mexico: *U.S. Bureau of Land Management*, Report no. 890127.

U.S. Bureau of Land Management, 1990, Bishop resource management plan and environmental impact statement: *U.S. Bureau of Land Management*, Report no. BLMCAES90001 1610, 273 p.

U.S. Bureau of Land Management, 1990, Rock Springs District wilderness study areas, Fremont, Lincoln, Sublette, and Sweetwater Counties, Wyoming: *U.S. Bureau of Land Management*, Report no. 900439, 324 p.

U.S. Bureau of Land Management, 1990, Uncompahgre Basin resource area, Montrose District, Colorado—Wilderness plan: *U.S. Bureau of Land Management*, Report no. 900175, 175 p.

U.S. Bureau of Land Management, 1991, Arizona Strip District resource management plan, Mohave and Coconino Counties, Arizona: *U.S. Bureau of Land Management*, Report no. 910055, 617 p.

U.S. Bureau of Land Management, 1991, resource management plan for public lands in the state of Nebraska: *U.S. Bureau of Land Management*, Report no. 910385, 144 p.

U.S. Bureau of Land Management, 1991, Safford District resource management plan, Arizona: *U.S. Bureau of Land Management*, Report no. 910314, 511 p.

U.S. Bureau of Land Management, 1991, Spokane resource management plan, Washington: *U.S. Bureau of Land Management*, Report no. 910374, 169 p.

U.S. Bureau of Land Management, 1991, Three Rivers resource management plan, Harney, Grant, Lake, and Malheur Counties, Oregon: *U.S. Bureau of Land Management*, Report no. 910335.

U.S. Bureau of Land Management, 1992, Proposed Spokane resource management plan amendment and final environmental impact statement: *U.S. Bureau of Land Management*, Report no. 920286, 181 p.

U.S. Bureau of Land Management, 1993, resource management plan and environmental impact statement for the Big Dry resource area, Miles City District, Montana: *U.S. Bureau of Land Management*, Report no. 930076, 382 p.

U.S. Bureau of Land Management, 1994, Burley District tiger beetle inventory: *U.S. Bureau of Land Management*, Report no. BLMTB942, 27 p.

U.S. Bureau of Land Management, 1994, Fort Wainwright, Yukon maneuver area, proposed resource management plan and final environmental impact statement, Fairbanks North Star Borough, Alaska: *U.S. Bureau of Land Management*, Report no. 940001, 124 p.

U.S. Bureau of Land Management, 1994, resource management plan and environmental impact statement for the Grass Creek resource area, Worland District, Wyoming: *U.S. Bureau of Land Management*, Report, 299 p.

U.S. Bureau of Land Management, 1994, Salem District resource management plan and environmental impact statement—Benton, Clackamas, Clatsop, Columbia, Lane, Lincoln, Linn, Marion, Multnomah, Polk, Tillamook, Washington, and Yamhill Counties, Oregon: *U.S. Bureau of Land Management*, Report no. 940461.

U.S. Bureau of Land Management, 1994, Stateline resource management plan, Clark and Nye Counties, Nevada (draft environmental impact statement and draft supplement): *U.S. Bureau of Land Management*, Report no. 940177.

U.S. Bureau of Land Management, 1996, resource management plan for the Green River resource area, Rock Springs, Wyoming: *U.S. Bureau of Land Management*, Report no. 960148.

U.S. Bureau of Land Management, 1996, Sweet Grass Hills resource management plan amendment, Liberty and Toole Counties, Montana: *U.S. Bureau of Land Management*, Report no. 960243, 179 p.

U.S. Bureau of Land Management, 1997, Proposed resource management plan/final environmental impact statement for the Roswell resource area, Roswell, New Mexico and proposed resource management plan amendment/final environmental impact statement for the Carlsbad resource area, Carlsbad, New Mexico. Volume 1: *U.S. Bureau of Land Management*, Report no. BLMNMPT970031610V1, 379 p.

U.S. Bureau of Land Management, 1997, Proposed resource management plan/Final environmental impact statement for the Roswell resource area, Roswell, New Mexico and proposed resource management plan amendment/final environmental impact statement for the Carlsbad resource area, Carlsbad, New Mexico. Volume 2: *U.S. Bureau of Land Management*, Report no. BLMNMPT970031610V2, 535 p.

U.S. Bureau of Land Management, 1997, Roswell and Carlsbad resource areas—Chaves, Curry, Debaca, Eddy, Guadalupe, Lea, Lincoln, Quay, and Roosevelt Counties, New Mexico: *U.S. Bureau of Land Management*, Report no. 970072.

U.S. Bureau of Land Management, 1998, Draft Caliente management framework plan amendment and environmental impact statement for management of desert tortoise habitat, Lincoln County, Nevada: *U.S. Bureau of Land Management*, Report no. 980190, 322 p.

U.S. Bureau of Land Management, 1998, Judith-Valley-Phillips resource management plan—Chouteau, Fergus, Judith Basin, Petroleum, Phillips, and Valley Counties, Montana (draft supplement to the Final environmental impact statement of September 1997): *U.S. Bureau of Land Management*, Report no. 980149, 51 p.

U.S. Bureau of Land Management, 1998, Socorro resource area resource management plan, New Mexico: *U.S. Bureau of Land Management*, Report no. 880397, 266 p.

U.S. Bureau of Land Management, 1999, Final environmental impact statement for management of desert tortoise habitat, Lincoln County, Nevada: *U.S. Bureau of Land Management*, Report no. 990201, 442 p.

U.S. Bureau of Land Management, 2000, Federal fluid minerals leasing and development, Otero and Sierra Counties, New Mexico: *U.S. Bureau of Land Management*, Report no. 000383, 521 p.

U.S. Bureau of Land Management, 2000, Riparian and aquatic habitat management in the Albuquerque Field Office—Bernalillo, Cibola, McKinley, Rio Arriba, Sandoval, Santa Fe, and Valencia Counties, New Mexico: *U.S. Bureau of Land Management*, Report no. 000304.

U.S. Bureau of Land Management, 2000, Riparian and aquatic habitat management in the Farmington Field Office—McKinley, Rio Arriba, San Juan, and Sandoval Counties, New Mexico: *U.S. Bureau of Land Management*, Report no. 000302.

U.S. Bureau of Land Management, 2000, Riparian and aquatic habitat management in the Las Cruces Field Office—Dona Ana, Grant, Hidaglo, and Luna Counties, New Mexico: *U.S. Bureau of Land Management*, Report no. 000303.

U.S. Bureau of Land Management, 2000, Riparian and aquatic habitat management in the Taos Field Office—Colfax, Harding, Los Alamos, Mora, Rio Arriba, San Miguel, Santa Fe, Taos, and Unison Counties, New Mexico: *U.S. Bureau of Land Management*, Report no. 00305.

U.S. Bureau of Land Management, 2001, National management strategy for motorized off-highway vehicle use on public lands: *U.S. Bureau of Land Management*, Report no. PB2001103 162, 58 p.

U.S. Bureau of Land Management, 2003, Farmington resource management plan, San Juan, McKinley, Rio Arriba, and Sandoval Counties, New Mexico: *U.S. Bureau of Land Management*, Report no. 030143.

U.S. Bureau of Land Management, 2004, Dillon resource management plan, Beaverhead and Madison Counties, Montana: *U.S. Bureau of Land Management*, Report no. 040153, 626 p.

U.S. Bureau of Land Management, 2004, Jack Morrow Hills coordinated activity plan, Sweetwater, Fremont, and Sublette Counties, Wyoming: *U.S. Bureau of Land Management*, Technical Report.

U.S. Bureau of Land Management, 2005, Newmont Mining Company, Emigrant project, Elko County, Nevada: *U.S. Bureau of Land Management*, Report no. 050117, 269 p.

U.S. Bureau of Land Management, 2006, Coeur d'Alene resource management plan, Idaho: *U.S. Bureau of Land Management*, Report no. 060002, 347 p.

U.S. Bureau of Land Management and U.S. Fish and Wildlife Service, 1992, Cokeville Meadows National Wildlife Refuge Proposal, Lincoln County, Wyoming: *U.S. Bureau of Land Management* and U.S. Fish and Wildlife Service, Report no. 920230, 123 p.

U.S. Bureau of Land Management and U.S. Forest Service, 1986, Proposed Coronado National Forest land and resource management plan, Arizona and New Mexico: *U.S. Bureau of Land Management* and U.S. Forest Service, Report no. 860307.

U.S. Bureau of Land Management and U.S. Forest Service, 1990, Proposed Cokeville Meadows National Wildlife Refuge, Lincoln County, Wyoming: *U.S. Bureau of Land Management* and U.S. Forest Service, Report no. 900388, 197 p.

U.S. Bureau of Land Management and U.S. Forest Service, 2004, North Fork of the South Platte and the South Platte Rivers wild and scenic river study report, Pike and San Isabel National Forests and Comanche and Cimarron National Grasslands, Douglas, Jefferson, Park, and Teller Counties, Colorado: *U.S. Bureau of Land Management* and U.S. Forest Service, Report no. 040037, 301 p.

U.S. Bureau of Reclamation, 2001, Pothole Reservoir resource management plan, Grant County, Washington: *U.S. Bureau of Reclamation, R*eport no. 010518, 567 p.

U.S. Department of Defense, 1999, Military training in the Marianas, Guam and the Commonwealth of the Northern Marianas: *U.S. Department of Defense,* Report no. 990187, 478 p.

U.S. Department of the Navy, 1973, Off-road vehicle use of Mirror Lake Naval Weapons Center, China Lake, California: *U.S. Department of the Navy, R*eport no. ELR73–141 1, 67 p.

U.S. Department of the Air Force, 1992, Defense evaluation support activity testing and evaluation program, Kirtland Air Force Base, New Mexico: *U.S. Department of the Air Force,* Report no. 920325, 487 p.

U.S. Department of the Air Force Tactical Air Command and U.S. Bureau of Land Management, 1986, Renewed withdrawal of Groom Mountain Range addition to Nellis Air Force bombing and gunnery range, Lincoln County, Nevada: U.S. Department of the Air Force, Tactical Air Command and *U.S. Bureau of Land Management,* Report no. 860476.

U.S. Federal Highway Administration, 1965, Design of roadside drainage channels: *U.S. Federal Highway Administration,* Report no. FHWAEPD86 103, 64 p.

U.S. Federal Highway Administration, 2005, US 33 Nelsonville bypass, city of Nelsonville, Hocking and Athens Counties, Ohio: *U.S. Federal Highway Administration,* Report no. 050282, 828 p.

U.S. Fish and Wildlife Service, 1985, Alaska Peninsula National Wildlife Refuge final comprehensive conservation plan, environmental impact statement, and wilderness review: *U.S. Fish and Wildlife Service,* Technical Report, 426 p.

U.S. Fish and Wildlife Service, 1985, Becharof National Wildlife Refuge comprehensive conservation plan and wilderness review, Alaska: *U.S. Fish and Wildlife Service,* Report no. 850193, 268 p.

U.S. Fish and Wildlife Service, 1986, Great Dismal Swamp National Wildlife Refuge master plan, Virginia and North Carolina: *U.S. Fish and Wildlife Service,* Report no. 860501, 245 p.

U.S. Forest Service, 1973, Aquarius Land Use Plan: *U.S. Forest Service,* Report no. USDAFSFESADM7411, 253 p.

U.S. Forest Service, 1973, Management of South Holston unit: *U.S. Forest Service,* Report no. USDAFSFESADM7350, 121 p.

U.S. Forest Service, 1973, Off-road vehicle policy, Hoosier National Forest, Indiana: *U.S. Forest Service,* Report no. USDAFS–DES(ADM)–73–51, 110 p.

U.S. Forest Service, 1973, Proposed off-road vehicle regulations and administrative instructions: *U.S. Forest Service,* Report no. USDAFS–FES(ADM)–73–49, 110 p.

U.S. Forest Service, 1973, Proposed regulations and administrative instructions relating to use of off-road vehicles on National Forest lands: *U.S. Forest Service,* Report no. USDAFS–DES(ADM)–73–49, 22 p.

U.S. Forest Service, 1973, Unit plan for management of the South Holston unit: *U.S. Forest Service,* Report no. USDAFS–DES(ADM)–73–50, 80 p.

U.S. Forest Service, 1985, Caribou National Forest and Curlew National Grassland land and resource management plan, Idaho, Utah, and Wyoming: *U.S. Forest Service,* Report no. 850437.

U.S. Forest Service, 1985, Forest land and resource management plan, Mark Twain National Forest, Missouri: *U.S. Forest Service,* Report no. 850118.

U.S. Forest Service, 1985, Kisatchie National Forest land and resource management plan, Louisiana: *U.S. Forest Service,* Report no. 850414.

U.S. Forest Service, 1985, Proposed land and resource management plan for the Medicine Bow National Forest and Thunder Basin National Grassland, Wyoming: *U.S. Forest Service,* Report no. 850523.

U.S. Forest Service, 1986, Chequamegon National Forest, Wisconsin land and resource management plan: *U.S. Forest Service,* Report no. 860325.

U.S. Forest Service, 1986, Cleveland National Forest land and resource management plan, Orange, Riverside, and San Diego Counties, California: *U.S. Forest Service,* Report no. 860209.

U.S. Forest Service, 1986, Croatan and Uwharrie National Forests land and resource management plan, North Carolina: *U.S. Department of Agriculture Forest Service,* Report no. 860218.

U.S. Forest Service, 1986, Land and resource management plan, Chippewa National Forest, Beltrami, Cass, and Itasca Counties, Minnesota: U.S. Forest Service, Report no. 860208.

U.S. Forest Service, 1986, Mark Twain National Forest land and resource management plan, Missouri: *U.S. Forest Service,* Report no. 860260.

U.S. Forest Service, 1986, National Forests in Alabama land and resource management plan: *U.S. Forest Service,* Report no. 860095.

U.S. Forest Service, 1986, Proposed Carson National Forest plan, Rio Arriba, Taos, Colfax, and Mora Counties, New Mexico: *U.S. Forest Service,* Report no. 860447.

U.S. Forest Service, 1986, Proposed land and resource management plan for Wayne National Forest, Ohio: *U.S. Forest Service,* Report no. 860426.

U.S. Forest Service, 1986, Proposed Santa Fe National Forest land and resource management plan, Mora, San Miguel, Santa Fe, Sandoval, Los Alamos, and Rio Arriba Counties, New Mexico: *U.S. Forest Service,* Report no. 860013.

U.S. Forest Service, 1987, Angeles National Forest land and resource management plan, Los Angeles, Ventura, and San Bernardino Counties, California: *U.S. Forest Service,* Report no. 870394.

U.S. Forest Service, 1987, Proposed Apache-Sitgreaves National Forests land and resource management plan, Apache, Coconino, Greenlee, and Navajo Counties, Arizona: *U.S. Forest Service,* Report no. 870387.

U.S. Forest Service, 1987, Proposed land and resource management plan for the Coconino Forest, Coconino, Gila, and Yavapai Counties, Arizona: *U.S. Forest Service,* Report no. 870288.

U.S. Forest Service, 1987, Santa Fe National Forest land management plan, Mora, San Miguel, Santa Fe, Sandoval, Los Alamos, and Rio Arriba Counties, New Mexico: *U.S. Forest Service,* Report no. 870299.

U.S. Forest Service, 1988, Proposed Kaibab National Forest land and resource management plan, Coconino, Yavapai, and Mohave Counties, Arizona: *U.S. Forest Service,* Report no. 880109.

U.S. Forest Service, 1988, Record of decision for USDA, Forest Service—Final environmental impact statement, Allegheny National Forest, land and resource management plan—Elk, Forest, McKean, and Warren Counties, Pennsylvania: *U.S. Forest Service,* Technical Report, 41 p.

U.S. Forest Service, 1989, San Bernardino National Forest land and resource management plan, San Bernardino and Riverside Counties, California: *U.S. Forest Service,* Report no. 890023.

U.S. Forest Service, 1990, Cherokee National Forest land and resource management plan, Tennessee, North Carolina, and Virginia (final supplement to the final environmental impact statement of April 1986): *U.S. Forest Service,* Report no. 900062, 62 p.

U.S. Forest Service, 1990, Umatilla National Forest land and resource management plan, Oregon and Washington: *U.S. Forest Service,* Report no. 900237.

U.S. Forest Service, 1991, Corral Mountain timber sale, San Juan National Forest, Archuleta County, Colorado: *U.S. Forest Service,* Report no. 910124, 59 p.

U.S. Forest Service, 1992, Martinez Creek timber sale, San Juan National Forest, Colorado: *U.S. Forest Service,* Report no. 920275, 75 p.

U.S. Forest Service, 1992, Norbeck Wildlife Preserve land management plan, Black Hills National Forest, Custer and Pennington Counties, South Dakota (final supplement to the environmental impact statement of July 1989): *U.S. Forest Service,* Report no. 920352, 92 p.

U.S. Forest Service, 1992, Rangeland ecosystem management in the Uinta National Forest, Provo, Utah: *U.S. Forest Service,* Report no. 920322, 189 p.

U.S. Forest Service, 1992, Sierra National Forest land and resource management plan, Fresno, Madera, and Mariposa Counties, California: *U.S. Forest Service,* Report no. 920105.

U.S. Forest Service, 1993, Six Rivers National Forest plan, Humboldt, Del Norte, Siskiyou, and Trinity Counties, California: *U.S. Forest Service,* Report no. 930340.

U.S. Forest Service, 1994, Oregon Dunes National Recreation Area management plan, Siuslaw National Forest—Coos, Douglas, and Lane Counties, Oregon: *U.S. Forest Service,* Report no. 940292.

U.S. Forest Service, 1994, *Revised land and resource management plan for the national forests and grasslands in Texas—Angelina, Fannin, Houston, Jasper, Montague, Montgomery, Nacogdoches, Newton, Sabine, San Augustine, San Jacinto, Shelby, Trinity, Walker, and Wise Counties*, Texas: U.S. Forest Service, Report no. 940385.

U.S. Forest Service, 1995, *Draft environmental impact statement, Fish Bate analysis area—North Fork Ranger District, Clearwater National Forest,* Clearwater County, Idaho: U.S. Forest Service, Technical Report.

U.S. Forest Service, 1995, East Fork Blacks Fork analysis area, Wasatch-Cache National Forest, Summit County, Utah (final supplement to the final environmental impact statement of July 1992): *U.S. Forest Service,* Report no. 950294, 49 p.

U.S. Forest Service, 1995, Mendocino National Forest land and resource management plan—Colusa, Glenn, Lake, Mendocino, Tehama, and Trinity Counties, California: *U.S. Forest Service,* Report no. 950337.

U.S. Forest Service, 1995, Snowy Trail re-route, Ventura County, California: *U.S. Forest Service,* Report no. 950117, 109 p.

U.S. Forest Service, 1996, Mendenhall Glacier Recreation Area management plan Revision, Chatham area, Juneau Ranger District, Tongass National Forest, Alaska: *U.S. Forest Service,* Report no. 960205, 279 p.

U.S. Forest Service, 1996, Proposed revised land and resource management plan for Francis Marion National Forest, Berkeley and Charleston Counties, South Carolina: *U.S. Forest Service,* Report no. 960158.

U.S. Forest Service, 1996, Snowy Trail re-route, Los Padres National Forest, Ventura County, California: *U.S. Forest Service,* Report no. 960524.

U.S. Forest Service, 2000, Silver Creek integrated resource project, Emmett Ranger District, Boise National Forest, Boise and Valley Counties, Idaho: *U.S. Forest Service,* Report no. 000241, 264 p.

U.S. Forest Service, 2001, Land and resource management plan, Uinta National Forest—Juab, Sanpete, Toole, Utah, and Wasatch Counties, Utah: *U.S. Forest Service,* Report no. 010142.

U.S. Forest Service, 2001, Starbucky restoration project, Red River Ranger District, Nez Perce National Forest, Idaho County, Idaho: *U.S. Forest Service,* Report no. 010274.

U.S. Forest Service, 2002, Final environmental impact statement amendment to the land and resource management plan—Management direction for acquired lands in southeastern Oklahoma, Ouachita National Forest, McCurtain County, Oklahoma: *U.S. Forest Service,* Technical Report.

U.S. Forest Service, 2002, Uncompahgre National Forest travel plan revision, Gunnison, Hinsdale, Mesa, Montrose, Ouray, San Juan, and San Miguel Counties, Colorado (final supplement to the final environmental impact statement of April 2000): *U.S. Forest Service,* Report no. 020131, 92 p.

U.S. Forest Service, 2002, West Gold Project, Sandpoint Ranger District, Idaho Panhandle National Forests, Bonner County, Idaho: *U.S. Forest Service,* Report no. 020484.

U.S. Forest Service, 2002, Whiskey Campo resource management project, Mountain Home Ranger District, Boise National Forest, Elmore County, Idaho: *U.S. Forest Service,* Report no. 020299, 151 p.

U.S. Forest Service, 2003, Big Bend Ridge vegetation management project and timber Sale, Ashton/Island Park Ranger District, Caribou-Targhee National Forest, Freemont County, Idaho: *U.S. Forest Service,* Report no. 030225.

U.S. Forest Service, 2003, Duck Creek–Swains access management project, Cedar City Ranger District, Dixie National Forest, Iron, Garfield, and Kane Counties, Utah: *U.S. Forest Service,* Report no. 030343, 49 p.

U.S. Forest Service, 2003, Upper Bear timber sale, Council Ranger District, Payette National Forest, Adams County, Idaho: *U.S. Forest Service,* Report no. 030319, 478 p.

U.S. Forest Service, 2004, Duck Creek fuels treatment analysis, Cedar City Ranger District, Dixie National Forest, Kane County, Utah: *U.S. Forest Service,* Report no. 040535, 240 p.

U.S. Forest Service, 2004, French Face ecosystem restoration, Ninemile Ranger District, Lolo National Forest, Montana: *U.S. Forest Service,* Report no. 040336, 184 p.

U.S. Forest Service, 2005, Dean project area, Bearlodge Ranger District, Black Hills National Forest, Crook County, Wyoming: *U.S. Forest Service,* Report no. 050212, 192 p.

U.S. Forest Service, 2005, Final environmental impact statement for the access designation on the Ocala National Forest, Lake, Marion, and Putnam Counties, Florida: *U.S. Forest Service,* Technical Report.

U.S. Forest Service, 2005, Gallatin National Forest travel management plan, Gallatin, Madison, Park, Meagher, Sweetgrass, and Carbon Counties, Montana: *U.S. Forest Service,* Report no. 050231.

U.S. Forest Service, 2005, Rocky Mountain Ranger District travel management plan, Rocky Mountain Ranger District, Lewis and Clark National Forest, Glacier, Pondera, Teton, and Lewis and Clark Counties, Montana: *U.S. Forest Service,* Report no. 050236, 386 p.

U.S. Forest Service, 2005, Woodrock project, Tongue Ranger District, Bighorn National Forest, Sheridan County, Wyoming: *U.S. Forest Service,* Report no. 050112, 201 p.

U.S. Forest Service, 2006, Hoosier National Forest land and resource management plan—Final environmental impact statement: *U.S. Forest Service,* Technical Report.

U.S. Forest Service, U.S. Bureau of Land Management, and U.S. Fish and Wildlife Service, 1997, George Washington National Forest land and resource management plan, Virginia and West Virginia (final supplement to the final environmental impact statement of January 1993): *U.S. Forest Service,* U.S. Bureau of Land Management, and U.S. Fish and Wildlife Service, Report no. 970046.

U.S. Immigration and Naturalization Service, 2004, U.S. Border Patrol activities within the border areas of the Tuscon and Yuma Sectors, Arizona (revision of the draft environmental impact statement of November 2002): *U.S. Immigration and Naturalization Service*, Report no. 040479, 487 p.

U.S. National Park Service, 1991, Big Cypress National Preserve, Collier, Monroe, and Dade Counties, Florida: *U.S. National Park Service,* Report no. 910386.

U.S. National Park Service, 1998, Redwood National and State Parks general management plan, Del Norte and Humboldt Counties, California: *U.S. National Park Service,* Report no. 980290, 474 p.

U.S. National Park Service, 1999, Big Cypress National Preserve, recreational off-road vehicle management plan—Collier, Dade, and Monroe Counties, Florida (draft supplement to the environmental impact statement of October 1991): *U.S. National Park Service,* Report no. 990285, 171 p.

U.S. National Park Service, 2000, Recreational Off-road vehicle management plan, Big Cypress National Preserve—Collier, Miami-Dade, and Monroe Counties, Florida (final supplement to the Final environmental impact statement of October 1991): *U.S. National Park Service,* Report no. 000275, 619 p.

U.S. Naval Facilities Engineering Command, 2003, Advanced amphibious assault vehicle, Marine Corps Base Camp Pendleton and San Clemente Range Complex, San Diego County, California: *U.S. Naval Facilities Engineering Command,* Report no. 030201.

Udevitz, M.S., Howard, C.A., Robel, R.J., and Curnutte, B., 1980, Lead contamination in insects and birds near an interstate highway, Kansas: *Environmental Entomology,* v. 9, no. 1, p. 35–36.

van der Zande, A.N., Keurs, W.J., and Van Der Weijden, W.J., 1980, The impact of roads on the densities of four bird species in an open field habitat—Evidence of a long-distance effect: *Biological Conservation,* v. 18, no. 4, p. 299–321.

Vos, C.C., and Chardon, J.P., 1998, Effects of habitat fragmentation and road density on the distribution pattern of the moor frog *Rana arvalis*: *Journal of Applied Ecology,* v. 35, no. 1, p. 44–56.

Warburton, M., Fisher, R., and Hathaway, S., 2004, Anza-Borrego Desert State Park amphibian survey, 2000-2002: *U.S. Geological Survey,* Report no. PB2005 105820, 44 p.

Ward, A.L., 1982, Mule deer behavior in relation to fencing and underpass on Interstate 80 in Wyoming: *Transportation Research Record,* v. 859, no. 8–13.

Watson, J.J., Kerley, G.I.H., and McLachlan, A., 1996, *Human activity and potential impacts on dune breeding birds in the Alexandria Coastal Dunefield: Landscape and Urban Planning,* v. 34, no. 3–4, p. 3 15–322.

Welch, R., Madden, M., and Doren, R.F., 1999, *Mapping the Everglades: Photogrammetric Engineering and Remote Sensing,* v. 65, no. 2, p. 163–170.

Wemple, B.C., Jones, J., and Grant, G., 1996, Channel network extension by logging roads in two basins, western Cascades, Oregon: *Water Resources Bulletin,* v. 32, no. 6, p. 1195–1207.

Wester, L., 1994, Weed management and the habitat protection of rare species—A case study of the endemic Hawaiian fern *Marsilea villosa*: *Biological Conservation,* v. 68, no. 1, p. 1–9.

Wheeler, A.P., Angermeier, P.L., and Rosenberger, A.E., 2005, Impacts of new highways and subsequent landscape urbanization on stream habitat and biota: *Reviews in Fisheries Science,* v. 13, no. 3, p. 141–164.

Whitten, K.R., and Cameron, R.D., 1983, Movements of collared caribou, *Rangifer tarandus*, in relation to petroleum development on the Arctic slope of Alaska: *Canadian Field-Naturalist,* v. 97, no. 2, p. 143–146.

Wiedemann, A.M., 1984, *Ecology of Pacific Northwest coastal sand dunes—A community profile:* Evergreen State College, Olympia, Washington, Report no. FWSOBS8404, 146 p.

Wilcox, B., and Murphy, D., 1985, Conservation strategy—The effects of fragmentation on extinction: *The American Naturalist,* v. 125, no. 6, p. 879–997.

Wild Utah Project, 1999, A literature review of the effects of off-road vehicles on desert biota, with emphasis on Utah BLM lands: *Wild Utah Project,* Salt Lake City, Utah, Report, 24 p.

Wilkins, K.T., 1982, Highways as barriers to rodent dispersal: *Southwestern Naturalist,* v. 27, no. 44, p. 459–460.

Williams, D.C., and Campbell, F., 1988, How the Bureau of Land Management designates and protects areas of critical environment concern—A status report, with a critical review by the Natural Resources Defense Council: *Natural Areas Journal,* v. 8, no. 4, p. 231–237.

Wilshire, H.G., 1983, Off-road vehicle recreation management policy for public lands in the United States—A case history: *Environmental Management,* v. 7, no. 6, p. 489–499.

Wilshire, H.G., Shipley, Susan, and Nakata, J.K., 1978, Impacts of off-road vehicles on vegetation: *Transactions of the North American Wildlife and Natural Resources Conference,* v. 43, p. 131–139.

Wisdom, M.J., Ager, A.A., Preisler, H.K., Cimon, N.J., and B.K. Johnson, 2005, Effects of off- road recreation on mule deer and elk, *in* Wisdom, M.J., ed., *The Starkey project—A synthesis of long-term studies of elk and mule deer: Lawrence, Kansas, Alliance Communications Group,* p. 67–80.

Wisdom, M.J., Cimon, N.J., Johnson, B.K., Garton, E.O., and Thomas, J.W., 2005, Spatial partitioning by mule deer and elk in relation to traffic, *in* Wisdom, M.J., ed., The Starkey project—*A synthesis of long-term studies of elk and mule deer: Lawrence, Kansas, Alliance Communications* Group, p. 53–66.

Wisdom, M.J., Rowland, M.M., Johnson, B.K., and Dick, B.L., 2005, Overview of the Starkey Project—Mule deer and elk research for management benefits, *in* Wisdom, M.J., ed., T*he Starkey Project—A synthesis of long-term studies of elk and mule deer:* Lawrence, Kansas, Alliance Communications Group, p. 17–28.

Wolcott, T.G., and Wolcott, D.L., 1984, Impact of off-road vehicles on macroinvertebrates of a mid-Atlantic beach: *Biological Conservation,* v. 29, no. 3, p. 2 17–240.

Wyatt, R.L., 1988, Implications of illegal off-road vehicle activity on the Cherokee Wildlife Management Area, Unicoi County, Tennessee: *Proceedings of the Annual Conference Southeastern Association of Fish and Wildlife Agencies,* v. 42, p. 533–539.

Yahner, R.H., 1988, Changes in wildlife communities near edges: *Conservation Biology,* v. 2, no. 4, p. 333–339.

Yahner, R.H., Morrell, T.E., and Rachael, J.S., 1989, Effects of edge contrast on depredation of artificial avian nests: *Journal of Wildlife Management,* v. 53, no. 4, p. 1135–1138.

Yanes, M., Velasco, J.M., and Suárez, F., 1995, Permeability of roads and railways to vertebrates—the importance of culverts: *Biological Conservation,* v. 71, no. 33, p. 2 17–222.

1.4. OHV Effects on Water Quality

Adams, E.S., 1975, Effects of lead and hydrocarbons from snowmobile exhaust on brook trout (*Salvelinus fontanlis*): *Transactions of the American Fisheries Society,* v. 104, no. 2, p. 363–373.

Al Awadhi, J.I., 2001, Impact of gravel quarrying on the desert environment of Kuwait: *Environmental Geology,* v. 41, no. 3–4, p. 365–371.

Anders, F.J., and Leatherman, S.P., 1987, Disturbance of beach sediment by off-road vehicles: *Environmental Geology and Water Sciences,* v. 9, no. 3, p. 183–189.

Assistant Secretary of Defense (Health and Environment), 1973, Proposed Department of Defense regulation on use of off-road vehicles: *U.S. Department of Defense,* Report no. ELR0511, 20 p.

Balades, J.D., Legret, M., and Madiec, H., 1995, Permeable pavements—Pollution management tools*: Water Science and Technology,* v. 32, no. 1, p. 49–56.

Bannister, E.N., 1979, Impact of road networks on southeastern Michigan lakeshore drainage: *Water Resources Research,* v. 15, no. 6, p. 15 15–1520.

Barrett, H., Cagney, J., Clark, R., Fogg, J., Gebhardt, K., Hansen, P.L., Mitchell, B., Prichard, D., and Tippy, D., 1995, Riparian area management—Process for assessing proper functioning condition: Denver, Colorado, *U.S. Bureau of Land Management,* Technical Reference 1737-9, BLM/SC/ST-93/003 + 1737 +REV95.

Beasley, G., and Kneale, P.E., 2003, Investigating the influence of heavy metals on macro-invertebrate assemblages using Partial Canonical Correspondence Analysis (pCCA): *Hydrology and Earth System Sciences,* v. 7, no. 2, p. 221–233.

Becking, R.W., and Hayes, G.L., 1970, Public land policy and the environment—Volume 3, part II (cont'd)—Environmental problems on the public lands—Case studies 9 through 17: *Rocky Mountain Center on Environment,* Denver, CO, Report no. PLLRC29–3, 436 p.

Belnap, Jayne, 1995, Surface disturbances—Their role in accelerating desertification: *Environmental Monitoring and Assessment,* v. 37, no. 1–3, p. 39–57.

Ben Brooks and Associates, Ironhorse Investors, Inc., Santa Fe Pacific Railroad Company, and U.S. Bureau of Land Management, 1998, *Hualapai Mountain Land Exchange,* Mohave County, Arizona: Report no. 980482, 175 p.

Bjornlie, D.D., and Garrott, R.A., 2001, Effects of winter road grooming on bison in Yellowstone National Park: *Journal of Wildlife Management,* v. 65, no. 3, p. 560–572.

Brabec, E., Schulte, S., and Richards, P.L., 2002, Impervious surfaces and water quality—A review of current literature and its implications for watershed planning: *Journal of Planning Literature,* v. 16, no. 4, p. 499–5 14.

Brodbeck, C.J., McDonald, T., Baier, J., and Taylor, S., 2004, Background water quality analysis from the Kentuck off-road vehicle trails, *in Proceedings, 2004 annual international meeting of the Society for Engineering in Agricultural, Food and Biological Systems,* Ottawa, Canada, August 1–4, 2004: St. Joseph, Michigan, American Society of Agricultural and Biological Engineers.

Brown, A.C., and McLachlan, A., 2002, Sandy shore ecosystems and the threats facing them— Some predictions for the year 2025: *Environmental Conservation,* v. 29, no. 1, p. 62–77.

Brown, G., and Porembski, S., 2000, Phytogenic hillocks and blow-outs as 'safe sites' for plants in an oil-contaminated area of northern Kuwait: *Environmental Conservation,* v. 27, no. 3, p. 242–249.

Brown, G., and Schoknecht, N., 2001, Off-road vehicles and vegetation patterning in a degraded desert ecosystem in Kuwait: *Journal of Arid Environments,* v. 49, no. 2, p. 413–427.

Brown, K.J., 1994, River-bed sedimentation caused by off-road vehicles at river fords in the Victorian Highlands, Australia: *Water Resources Bulletin,* v. 30, no. 2, p. 239–250.

Buckley, R., and King, N., 2003, Visitor-impact data in a land-management context, *in* Buckley, R.C., Pickering, D., and Weaver, D.B., eds., *Nature-based tourism, environment and land management:* Wallingford, England, CABI Publishing, p. 89-99.

California State Water Resources Control Board, 1980, Lake Tahoe Basin water quality plan, final plan: Sacramento, California, *California State Water Resources Control Board,* Technical Report.

Coats, R.N., and Miller, T.O., 1981, Cumulative silvicultural impacts on watersheds—A hydrologic and regulatory dilemma: *Environmental Management,* v. 5, no. 2, p. 147–160.

Collins, E., O'Farrell, T.P., and Rhoads, W., 1982, *Annotated bibliography for biologic overview for the Nevada Nuclear Waste Storage investigations,* Nevada Test Site, Nye County, Nevada: Goleta, California, EG and G, Inc., Report no. EGG1 1832419, 48 p.

Collins, E., O'Farrell, T.P., and Rhoads, W., 1982, *Biologic overview for the Nevada nuclear waste storage investigations,* Nevada Test Site, Nye County, Nevada: Goleta, California, EG and G, Inc., Report no. EGG1 1832460, 55 p.

Colwill, D.M., Peters, C.J., and Perry, R., 1984, Water quality of motorway runoff: Crowthorne, England, *Transport and Road Research Laboratory*, Report no. TRRL/SR–823, 28 p.

Davidson, E.D., and Fox, M., 1974, *Effects of off-road motorcycle activity on Mojave Desert vegetation and soil:* Madroño, v. 22, no. 8, p. 381–390.

Dawson, K.J., 1985, Natural area planning for recreational use transition: *Landscape Planning,* v. 12, no. 2, p. 111–123.

Deletic, A., Ashley, R., and Rest, D., 2000, Modelling input of fine granular sediment into drainage systems via gully-pots: *Water Research,* v. 34, no. 15, p. 3836–3844.

Eckert, R.E.J., Wood, M.K., Blackburn, W.H., and Peterson, F.F., 1979, Impacts of off-road vehicles on infiltration and sediment production of two desert soils: *Journal of Range Management,* v. 32, no. 5, p. 394–397.

Fang, S., Wente, S., Gertner, G.Z., Wang, G., and Anderson, A., 2002, Uncertainty analysis of predicted disturbance from off-road vehicular traffic in complex landscapes at Fort Hood: *Environmental Management,* v. 30, no. 2, p. 199–208.

Furniss, M.J., Flanagan, S.A., and McFadin, B.A., 2000, Hydrologically connected roads—An indicator of the influence of roads on chronic sedimentation, surface water hydrology, and exposure to toxic chemicals: *U.S. Forest Service,* Stream Systems Technology Center, Rocky Mountain Research Station, Technical Report, 4 p.

Glass, R.L., 1986, Hydrologic conditions in Connors Bog area, Anchorage, Alaska: *U.S. Geological Survey,* Technical Report, 23 p.

Greene, A.F.C., 1975, The need for cooperative approaches to fish and wildlife management: *Transactions of the North American Wildlife and Natural Resources Conference,* v. 40, p. 133–141.

Hairsine, P.B., Croke, J.C., Mathews, H., Fogarty, P., and Mockler, S.P., 2002, Modelling plumes of overland flow from logging tracks: *Hydrological Processes,* v. 16, no. 12, p. 23 11–2327.

Hamilton, L.J., 2002, A study of the effects of ORV stream crossings on water quality of two streams located in the Angelina National Forest, Texas—A physicochemical and benthic macroinvertebrate analysis: *Masters Abstracts International,* v. 40, no. 3, p. 668.

Hansen, R.P., Hillhouse, W.A., Willard, B.E., and Burke, H.D., 1970, Public land policy and the environment. Volume 2. Part II—*Environmental problems on the public lands—Summary statement and case studies 1 through 8:* Denver, Colorado, Rocky Mountain Center on Environment, Report no. PLLRC29–2, 409 p.

Haupt, H.F., 1959, Road and slope characteristics affecting sediment movement from logging roads: *Journal of Forestry,* v. 57, no. 5, p. 329–339.

Hayes, D.L., 1991, The all-American canal lining project—A catalyst for rational and comprehensive groundwater management on the United States-Mexico border: *Natural Resources Journal,* v. 31, no. 4, p. 803–827.

Hillel, D., and Tadmor, N., 1962, Water regime and vegetation in central Negev Highlands of Israel: *Ecology,* v. 43, no. 1, p. 33–41.

Hinckley, B.S., Iverson, R.M., and Hallet, B., 1983, Accelerated water erosion in ORV off-road vehicle-use areas, *in* Webb, R.H., and Wilshire, H.G., eds., *Environmental effects of off-road vehicles—Impacts and management in arid regions:* New York, Springer-Verlag, p. 8 1–96.

Hutton, M., and Symon, C., 1986, The quantities of cadmium, lead, mercury and arsenic entering the U.K. environment from human activities: *Science of the Total Environment,* v. 57, p. 129–150.

Iverson, R.M., 1980, Processes of accelerated pluvial erosion on desert hillslopes modified by vehicular traffic: *Earth Surface Processes,* v. 5, p. 369-3 88.

Johnson, R.R., Mills, G.S., and Carothers, S.W., 1990, Creation and restoration of riparian habitat in southwestern arid and semi-arid regions, *in* Kusler, J.A., and Kentula, M.E., eds., *Wetland creation and restoration—The status of the science:* Covelo, California, Island Press, p. 351–366.

Johnston, F.M., and Johnston, S.W., 2004, Impacts of road disturbance on soil properties and on exotic plant occurrence in subalpine areas of the Australian Alps: *Arctic, Antarctic, and Alpine Research,* v. 36, no. 2, p. 201–207.

Katz, M., Legore, R.S., Weitkamp, D., Cummins, J.M., and Anderson, D., 1972, Effects on freshwater fish: *Journal of the Water Pollution Control Federation,* v. 44, no. 6, p. 1226–1250.

Knott, J.M., 1978, Reconnaissance assessment of erosion and sedimentation in the Canada De Los Alamos Basin, Los Angeles and Ventura Counties, California: *U.S. Geological Survey,* Open File Report no. 78–873, 49 p.

Kondolf, G.M., Piegay, H., and Landon, N., 2002, Channel response to increased and decreased bedload supply from land use change—Contrasts between two catchments: *Geomorphology,* v. 45, no. 1–2, p. 35–51.

Koppel, W., 1988, Dynamic impact on soil structure due to traffic of off-road vehicles, *in* Drescher, J., Horn, R., and Boodt, M.D., eds., *Impact of water and external forces on soil structure—Selected papers of the 1st Workshop on Soilphysics and Soilmechanics*: Cremlingen-Destedt, Germany, Catena Verlag, p. 113–122.

Kurczerski, F.E., 2000, History of white pine (*Pinus strobus*)/oak (*Quercus* spp.) savanna in southern Ontario, with particular reference to the biogeography and status of the antenna-waving wasp, *Tachysphex pechumani* (Hymenoptera: Sphecidae): *Canadian Field-Naturalist,* v. 114, no. 1, p. 1–20.

Lewis, O., 1969, Arid lands and their future, *in* Bender, G.L., ed., Future environments of arid regions of the southwest, contribution no. 12, American Association for the Advancement of Science Committee on Desert and Arid Zone Research symposium, Colorado Springs, Colorado, May 7-10, 1969: Washington, D.C., *American Association for the Advancement of Science, Committee on Desert and Arid Zone Research,* p. 33–3 8.

Lovich, J.E., and Bainbridge, D., 1999, Anthropogenic degradation of the southern California desert ecosystem and prospects for natural recovery and restoration: *Environmental Management,* v. 24, no. 3, p. 309–326.

Melvin, S.M., Griffin, C.R., and MacIvor, L.H., 1991, Recovery strategies for piping plovers in managed coastal landscapes: *Coastal Management,* v. 19, no. 1, p. 2 1–34.

Mize, R., Evans, R.E., MacRoberts, B.R., MacRoberts, M.H., and Rudolph, D.C., 2005, Restoration of pitcher plant bogs in eastern Texas, USA: *Natural Areas Journal,* v. 25, no. 2, p. 197–201.

Morgan, M.T., 1993, Nutrition is the key to plight of an ancient survivor, the desert tortoise: *Smithsonian Institution Research Reports,* v. 74, p. 4.

Nicola, N.C., and Lovich, J.E., 2000, Preliminary observations of the behavior of male, flat-tailed horned lizards before and after an off-highway vehicle race in California: *California Fish and Game,* v. 86, p. 208–2 12.

O'Farrell, T.P., and Kato, T.T., 1987, *Biological assessment of the effects of petroleum production activities, Naval Petroleum Reserves in California, on the endangered giant kangaroo rat, Dipodomys ingens:* Goleta, California, EG and G Energy Measurements, Inc., Report no. EGG102822183, 34 p.

Okello, J.A., 1991, A review of soil strength measurement techniques for prediction of terrain vehicle performance: *Journal of Agricultural Engineering Research,* v. 50, no. 2, p. 129–155.

Paschka, M.G., Ghosh, R.S., and Dzombak, D.A., 1999, Potential water-quality effects from iron cyanide anticaking agents in road salt: *Water Environment Research,* v. 71, no. 6, p. 1235–1239.

Perdikaki, K., and Mason, C.F., 1999, Impact of road run-off on receiving streams in eastern England: *Water Research,* v. 33, no. 7, p. 1627–1633.

Persico, L.P., Nichols, K.K., and Bierman, P.R., 2005, Tracking painted pebbles—Short-term rates of sediment movement on four Mojave Desert piedmont surfaces: *Water Resources Research,* v. 41, no. 7, p. W07004.07001–W07004.07015.

Piehl, B.T., Beschta, R.L., and Pyles, M.R., 1988, Ditch-relief culverts and low-volume forest roads in the Oregon Coast Range: *Northwest Science,* v. 62, no. 3, p. 9 1–98.

Rickard, W.H., 1988, *Natural vegetation at the proposed reference repository location in southeastern Washington:* Richland, Washington, Battelle Pacific Northwest Labs, Report no. PNL–6402, 27 p.

Rinne, J.N., Rickel, B., and Hendrickson, D., 1980, A new Gila topminnow locality in southern Arizona: Fort Collins, Colorado, *U.S. Forest Service,* Rocky Mountain Forest and Range Experiment Station, Report no. RM-3 82, 4 p.

Robinson, E., 1999, A geology of the road: *Geology Today,* v. 15, no. 3, p. 110–113.

Roy, A.H., Rosemond, A.D., Leigh, D.S., Paul, M.J., and Wallace, J.B., 2003, Habitat-specific responses of stream insects to land cover disturbance—Biological consequences and monitoring implications: *Journal of the North American Benthological Society,* v. 22, no. 2, p. 292–307.

Samaras, Z., and Zierock, K.H., 1995, Off-road vehicles—A comparison of emissions with those from road transport: *Science of the Total Environment,* v. 169, p. 249–253.

Schemnitz, S.D., and Schortemeyer, J.D., 1972, *The influence of vehicles on Florida Everglades vegetation:* Tallahassee, Florida, Florida State Game and Fresh Water Fish Commission, Report no. DISFEP–74–3 1, 65 p.

Schultink, G., 1977, *Impact analysis of off-road-vehicle use on vegetation in the Grand Mere Dune environment: East Lansing,* Michigan, Michigan State University, Report no. NASACR155764, 10 p.

Severinghaus, W.D., Riggins, R.E., and Goran, W.D., 1979, Effects of tracked vehicle activity on terrestrial mammals and birds at Fort Knox, KY (special report): *Transactions of the Kentucky Academy of Science,* v. 41, no. 1–2, p. 15–26.

Snyder, C.T., Frickel, D.G., Hadley, R.F., and Miller, R.F., 1976, Effects of off-road vehicle use on the hydrology and landscape of arid environments in central and southern California: *U.S. Geological Survey,* Water Resources Investigations Report no. 76–99, 52 p.

Sparrow, S.D., Wooding, F.J., and Whiting, E.H., 1978, Effects of off-road vehicle traffic on soils and vegetation in the Denali Highway region of Alaska: *Journal of Soil and Water Conservation,* v. 33, no. 1, p. 20–27.

Steiner, A.J., and Leatherman, S.P., 1981, Recreational impacts on the distribution of ghost crabs *Ocypode quadrata* Fab*: Biological Conservation,* v. 20, no. 2, p. 111–122.

Tunstall, B.R., and Reece, P.H., 1989, Environmental assessment of the Sunset and Big Desert lands, northwest Victoria, Australia: Clayton South, Australia, Australia Commonwealth Scientific and Industrial Research Organization, *Division of Water Resources Divisional Report,* Report, 1–85 p.

Turton, S.M., 2005, Managing environmental impacts of recreation and tourism in rainforests of the wet tropics of Queensland World Heritage Area: *Geographical Research,* v. 43, no. 2, p. 140–151.

Tuttle, M., and Griggs, G., 1985, Accelerated soil erosion at three state vehicular recreation Areas, central and southern California, *Erosion control— A challenge in our time, proceedings of the 16th annual International Erosion Control Association,* February 2 1–22, 1985, San Francisco, California: San Francisco, California, International Erosion Control Association, p. 105-115.

Tuttle, M., and Griggs, G., 1987, Soil erosion and management recommendations at three state vehicular recreation Areas, California*: Environmental Geology and Water Sciences,* v. 10, no. 2, p. 111–123.

Twiss, R., Sidener, J., Bingham, G., Burke, J.E., and Hall, C.H., 1980, *Potential impacts of geothermal development on outdoor recreational use of the Salton Sea: Livermore,* California, University of California at Livermore, Technical Report, 61 p.

U.S. Air Force, 1992, Defense evaluation support activity testing and evaluation program, Kirtland Air Force Base, New Mexico*: U.S. Air Force,* Report no. 920325, 487 p.

U.S. Air Force Tactical Air Command and U.S. Bureau of Land Management, 1986, Renewed withdrawal of Groom Mountain Range addition to Nellis Air Force bombing and gunnery range, Lincoln County, Nevada: *U.S. Department of the Air Force,* Tactical Air Command and U.S. Bureau of Land Management, Report no. 860476.

U.S. Army Corps of Engineers, 1994, Stationing of mechanized or armored combat forces at Fort Lewis, Thurston and Pierce Counties, Washington: *U.S. Army Corps of Engineers,* Report no. 940045.

U.S. Army Corps of Engineers, 2001, Fort Bliss mission and master plan—Dona Ana and Otero Counties, New Mexico, and El Paso County, Texas: *U.S. Army Corps of Engineers,* Report no. 010081.

U.S. Army Corps of Engineers, and Oklahoma Department of Wildlife Conservation, 1986, Candy Lake, Candy Creek, Osage County, Oklahoma (supplemental information report to the final environmental impact statement of January 1975)*: U.S. Army Corps of Engineers,* Report no. 860246, 4 p.

U.S. Bureau of Land Management, 1984, Book Cliffs resource management plan, Utah: *U.S. Bureau of Land Management,* Report no. 840541, 534 p.

U.S. Bureau of Land Management, 1984, Camping on the public lands: *U.S. Bureau of Land Management,* Report no. PB 85204907, 5 p.

U.S. Bureau of Land Management, 1984, Proposed Monument resource management plan, Idaho*: U.S. Bureau of Land Management,* Report no. 840584, 419 p.

U.S. Bureau of Land Management, 1985, Grand Junction resource area—Resource management plan and environmental impact statement: *U.S. Bureau of Land Management,* Report, 171 p.

U.S. Bureau of Land Management, 1985, Lower Gila South resource management plan, La Paz, Maricopa, Pima, Pinal, and Yuma Counties, Arizona: *U.S. Bureau of Land Management,* Report no. 850347, 909 p.

U.S. Bureau of Land Management, 1985, Medicine Lodge proposed resource management plan, Bingham, Bonneville, Butte, Clark, Fremont, Jefferson, Madison, and Teton Counties, Idaho: *U.S. Bureau of Land Management,* Report no. 850237.

U.S. Bureau of Land Management, 1985, resource management plan for the Kemmerer resource area, Lincoln, Sublette, Sweetwater, and Uinta Counties, Wyoming: *U.S. Bureau of Land Management,* Report no. 850532.

U.S. Bureau of Land Management, 1985, resource management plan for the Walker planning Area, Nevada: *U.S. Bureau of Land Management,* Report no. 850274.

U.S. Bureau of Land Management, 1985, White Sands resource area resource management plan, New Mexico: *U.S. Bureau of Land Management,* Report no. 850387.

U.S. Bureau of Land Management, 1986, Baker resource management plan, Baker County, Oregon: *U.S. Bureau of Land Management,* Report no. 860150, 129 p.

U.S. Bureau of Land Management, 1986, Big Lost/Pahsimeroi Wilderness, Idaho: *U.S. Bureau of Land Management,* Report no. 860399, 125 p.

U.S. Bureau of Land Management, 1986, Eastern San Diego County planning unit, California— Preliminary wilderness recommendations: *U.S. Bureau of Land Management,* Report no. 860388, 195 p.

U.S. Bureau of Land Management, 1986, Lander resource management plan, Lander, Wyoming: *U.S. Bureau of Land Management,* Report no. 860449.

U.S. Bureau of Land Management, 1986, Little Snake resource management plan, Moffat, Rio Blanco, and Routt Counties, Colorado: *U.S. Bureau of Land Management,* Report no. 860029.

U.S. Bureau of Land Management, 1986, Proposed Baker resource management plan, Vale District, Oregon: *U.S. Bureau of Land Management,* Report no. 860396.

U.S. Bureau of Land Management, 1986, resource management plan for the Carlsbad resource area, Eddy, Lea, and Chaves Counties, New Mexico: *U.S. Bureau of Land Management,* Report no. 860101, 390 p.

U.S. Bureau of Land Management, 1986, Utah statewide wilderness environmental impact statement: *U.S. Bureau of Land Management,* Report no. 860036.

U.S. Bureau of Land Management, 1986, Wilderness Suitability EIS—Grass Creek and Cody resource areas, Bighorn Basin, Wyoming (draft supplement for the Owl Creek Wilderness Study Area): *U.S. Bureau of Land Management,* Report no. 860168, 40 p.

U.S. Bureau of Land Management, 1987, Medicine Bow-Divide resource areas resource management plan, Rawlins District, Wyoming: *U.S. Bureau of Land Management,* Report no. 870212.

U.S. Bureau of Land Management, 1987, North Dakota resource management plan, Dunn and Bowman Counties, North Dakota: *U.S. Bureau of Land Management,* Report no. 870236, 199 p.

U.S. Bureau of Land Management, 1987, Pinedale resource area resource management plan, Sublette and Lincoln Counties, Wyoming: *U.S. Bureau of Land Management,* Report no. 870077, 347 p.

U.S. Bureau of Land Management, 1987, Pocatello resource management plan, Idaho: *U.S. Bureau of Land Management,* Report no. 870333, 132 p.

U.S. Bureau of Land Management, 1987, Proposed Lemhi resource management plan, Lemhi County, Idaho—Eighteenmile Wilderness: *U.S. Bureau of Land Management,* Report no. 870437, 178 p.

U.S. Bureau of Land Management, 1987, Proposed Medicine Lodge, Idaho resource management plan—Wilderness recommendations: *U.S. Bureau of Land Management,* Report no. 870436, 92 p.

U.S. Bureau of Land Management, 1987, Proposed resource management plan for the San Juan resource area, Moab District, Utah: *U.S. Bureau of Land Management,* Report no. 870440.

U.S. Bureau of Land Management, 1987, Proposed utility corridor resource management plan, Alaska: *U.S. Bureau of Land Management,* Report no. 870291, 281 p.

U.S. Bureau of Land Management, 1987, Washakie resource area resource management plan, portions of Big Horn, Hot Springs, and Washakie Counties, Wyoming: *U.S. Bureau of Land Management,* Report no. 870395.

U.S. Bureau of Land Management, 1988, Brothers/Lapine resource management plan, Oregon: *U.S. Bureau of Land Management,* Report no. 880365.

U.S. Bureau of Land Management, 1988, Challis resource area proposed resource management plan, upper Columbia-Salmon Clearwater Districts, Custer and Lemhi Counties, Idaho: *U.S. Bureau of Land Management,* Report no. 980508.

U.S. Bureau of Land Management, 1988, Cody resource area resource management plan, Big Horn and Park Counties, Wyoming: *U.S. Bureau of Land Management,* Report no. 880315.

U.S. Bureau of Land Management, 1988, Fort Greely national maneuver area and Fort Greely air drop done, Alaska—Resource management plan: *U.S. Bureau of Land Management,* Report no. 880276, 138 p.

U.S. Bureau of Land Management, 1988, Fort Wainwright maneuver area, Fairbanks North Star Borough, Alaska: *U.S. Bureau of Land Management,* Report no. 880277, 138 p.

U.S. Bureau of Land Management, 1988, Medicine Bow-Divide resource areas (now Great Divide resource area) resource management plan, Rawlins District, Wyoming: *U.S. Bureau of Land Management,* Report no. 880325.

U.S. Bureau of Land Management, 1988, Pony Express resource management plan, Toole, Utah, and Salt Lake Counties, Utah: *U.S. Bureau of Land Management,* Report no. 880310, 147 p.

U.S. Bureau of Land Management, 1988, Proposed Las Vegas resource management plan and final environmental impact statement, Clark and Nye Counties, Nevada: *U.S. Bureau of Land Management,* Report no. 980207.

U.S. Bureau of Land Management, 1988, Uncompahgre Basin resource management plan, Colorado: *U.S. Bureau of Land Management,* Report no. 880345.

U.S. Bureau of Land Management, 1988, West Hiline resource management plan, northcentral Montana: *U.S. Bureau of Land Management,* Report no. 880200, 306 p.

U.S. Bureau of Land Management, 1988, Wilderness study areas in the Rock Springs District, Fremont, Lincoln, Sublette, and Sweetwater Counties, Wyoming (revised Draft environmental impact statement): *U.S. Bureau of Land Management,* Report no. 880359, 175 p.

U.S. Bureau of Land Management, 1989, Dixie resource area management plan, Washington County, Utah: *U.S. Bureau of Land Management,* Report no. 890290, 250 p.

U.S. Bureau of Land Management, 1989, Phoenix resource management plan, Arizona: *U.S. Bureau of Land Management,* Report no. 890012, 240 p.

U.S. Bureau of Land Management, 1989, San Pedro River riparian management plan for the San Pedro River EIS Area, Cochise County, Arizona: *U.S. Bureau of Land Management,* Report no. 890152, 381 p.

U.S. Bureau of Land Management, 1989, San Rafael resource management plan, Emery County, Utah: *U.S. Bureau of Land Management,* Report no. 890240.

U.S. Bureau of Land Management, 1989, Utility corridor proposed resource management plan, Alaska: *U.S. Bureau of Land Management,* Report no. 890326, 281 p.

U.S. Bureau of Land Management, 1990, Bishop resource management plan and environmental impact statement: *U.S. Bureau of Land Management,* Report no. BLMCAES90001 1610, 273 p.

U.S. Bureau of Land Management, 1990, Rock Springs District wilderness study areas, Fremont, Lincoln, Sublette, and Sweetwater Counties, Wyoming: *U.S. Bureau of Land Management,* Report no. 900439, 324 p.

U.S. Bureau of Land Management, 1991, Arizona Strip District resource management plan, Mohave and Coconino Counties, Arizona: *U.S. Bureau of Land Management,* Report no. 910055, 617 p.

U.S. Bureau of Land Management, 1991, resource management plan for public lands in the state of Nebraska: *U.S. Bureau of Land Management,* Report no. 910385, 144 p.

U.S. Bureau of Land Management, 1991, Safford District resource management plan, Arizona: *U.S. Bureau of Land Management,* Report no. 910314, 511 p.

U.S. Bureau of Land Management, 1991, Three Rivers resource management plan, Harney, Grant, Lake, and Malheur Counties, Oregon: *U.S. Bureau of Land Management,* Report no. 910335.

U.S. Bureau of Land Management, 1994, Fort Wainwright, Yukon maneuver area, proposed resource management plan and final environmental impact statement, Fairbanks North Star Borough, Alaska: *U.S. Bureau of Land Management,* Report no. 940001, 124 p.

U.S. Bureau of Land Management, 1994, resource management plan and environmental impact statement for the Grass Creek resource area, Worland District, Wyoming: *U.S. Bureau of Land Management,* Report, 299 p.

U.S. Bureau of Land Management, 1994, Salem District resource management plan and environmental impact statement—Benton, Clackamas, Clatsop, Columbia, Lane, Lincoln, Linn, Marion, Multnomah, Polk, Tillamook, Washington, and Yamhill Counties, Oregon: *U.S. Bureau of Land Management,* Report no. 940461.

U.S. Bureau of Land Management, 1996, resource management plan for the Green River resource area, Rock Springs, Wyoming: *U.S. Bureau of Land Management,* Report no. 960148.

U.S. Bureau of Land Management, 1996, Sweet Grass Hills resource management plan amendment, Liberty and Toole Counties, Montana: *U.S. Bureau of Land Management,* Report no. 960243, 179 p.

U.S. Bureau of Land Management, 1997, Proposed resource management plan/final environmental impact statement for the Roswell resource area, Roswell, New Mexico and proposed resource management plan Amendment/final environmental impact statement for the Carlsbad resource area, Carlsbad, New Mexico. Volume 1: *U.S. Bureau of Land Management,* Report no. BLMNMPT970031610V1, 379 p.

U.S. Bureau of Land Management, 1997, Proposed resource management plan/final environmental impact statement for the Roswell resource area, Roswell, New Mexico and proposed resource management plan amendment/final environmental impact statement for the Carlsbad resource area, Carlsbad, New Mexico. Volume 2: *U.S. Bureau of Land Management,* Report no. BLMNMPT970031610V2, 535 p.

U.S. Bureau of Land Management, 1998, Judith-Valley-Phillips resource management plan—Chouteau, Fergus, Judith Basin, Petroleum, Phillips, and Valley Counties, Montana (draft supplement to the final environmental impact statement of September 1997): *U.S. Department of Interior Bureau of Land Management,* Report no. 980149, 51 p.

U.S. Bureau of Land Management, 1998, Socorro resource area resource management plan, New Mexico: *U.S. Department of Interior Bureau of Land Management,* Report no. 880397, 266 p.

U.S. Bureau of Land Management, 2000, Federal fluid minerals leasing and development, Otero and Sierra Counties, New Mexico: *U.S. Department of Interior Bureau of Land Management,* Report no. 000383, 521 p.

U.S. Bureau of Land Management, 2000, Pinedale Anticline Oil and Gas Exploration and Development Project, Sublette County, Wyoming: *U.S. Department of Interior Bureau of Land Management*, Report no. 000159, 301 p.

U.S. Bureau of Land Management, 2000, Riparian and aquatic habitat management in the Albuquerque Field Office—Bernalillo, Cibola, McKinley, Rio Arriba, Sandoval, Santa Fe, and Valencia Counties, New Mexico: *U.S. Bureau of Land Management,* Report no. 000304.

U.S. Bureau of Land Management, 2000, Riparian and aquatic habitat management in the Farmington Field Office—McKinley, Rio Arriba, San Juan, and Sandoval Counties, New Mexico: *U.S. Bureau of Land Management,* Report no. 000302.

U.S. Bureau of Land Management, 2000, Riparian and aquatic habitat management in the Las Cruces Field Office—Dona Ana, Grant, Hidaglo, and Luna Counties, New Mexico: *U.S. Bureau of Land Management,* Report no. 000303.

U.S. Bureau of Land Management, 2000, Riparian and aquatic habitat management in the Taos Field Office—Colfax, Harding, Los Alamos, Mora, Rio Arriba, San Miguel, Santa Fe, Taos, and Unison Counties, New Mexico: *U.S. Bureau of Land Management,* Report no. 00305.

U.S. Bureau of Land Management, 2001, National management strategy for motorized off-highway vehicle use on public lands: *U.S. Bureau of Land Management,* Report no. PB2001103 162, 58 p.

U.S. Bureau of Land Management, 2004, Dillon resource management plan, Beaverhead and Madison Counties, Montana: *U.S. Bureau of Land Management,* Report no. 040153, 626 p.

U.S. Bureau of Land Management, 2005, Newmont Mining Company, Emigrant project, Elko County, Nevada: *U.S. Bureau of Land Management,* Report no. 050117, 269 p.

U.S. Bureau of Land Management and U.S. Fish and Wildlife Service, 1992, Cokeville Meadows National Wildlife Refuge proposal, Lincoln County, Wyoming: U.S. Bureau of Land Management and U.S. Fish and Wildlife Service, Report no. 920230, 123 p.

U.S. Bureau of Land Management and U.S. Forest Service, 1986, Proposed Coronado National Forest land and resource management plan, Arizona and New Mexico: U.S. Bureau of Land Management and U.S. Forest Service, Report no. 860307.

U.S. Bureau of Land Management and U.S. Forest Service, 1990, Proposed Cokeville Meadows National Wildlife Refuge, Lincoln County, Wyoming: U.S. Bureau of Land Management and U.S. Forest Service, Report no. 900388, 197 p.

U.S. Bureau of Land Management and U.S. Forest Service, 2004, North Fork of the South Platte and the South Platte Rivers wild and scenic river study report, Pike and San Isabel National Forests and Comanche and Cimarron National Grasslands, Douglas, Jefferson, Park, and Teller Counties, Colorado: U.S. Bureau of Land Management and U.S. Forest Service, Report no. 040037, 301 p.

U.S. Bureau of Reclamation, 2001, Pothole Reservoir resource management plan, Grant County, Washington: U.S. Bureau of Reclamation, Report no. 010518, 567 p.

U.S. Federal Highway Administration, 2005, US 33 Nelsonville bypass, city of Nelsonville, Hocking and Athens Counties, Ohio: U.S. Department of Transportation, Federal Highway Administration, Report no. 050282, 828 p.

U.S. Fish and Wildlife Service, 1986, Great Dismal Swamp National Wildlife Refuge master plan, Virginia and North Carolina: U.S. Fish and Wildlife Service, Report no. 860501, 245 p.

U.S. Forest Service, 1973, Off-road vehicle policy, Hoosier National Forest, Indiana: U.S. Forest Service, Report no. USDAFS–DES(ADM)–73–51, 110 p.

U.S. Forest Service, 1973, Proposed off-road vehicle regulations and administrative instructions: U.S. Forest Service, Report no. USDAFS–FES(ADM)–73–49, 110 p.

U.S. Forest Service, 1973, Proposed regulations and administrative instructions relating to use of off-road vehicles on National Forest lands: U.S. Forest Service, Report no. USDAFS–DES(ADM)–73–49, 22 p.

U.S. Forest Service, 1985, Caribou National Forest and Curlew National Grassland land and resource management plan, Idaho, Utah, and Wyoming: U.S. Forest Service, Report no. 850437.

U.S. Forest Service, 1985, Kisatchie National Forest land and resource management plan, Louisiana: U.S. Forest Service, Report no. 850414.

U.S. Forest Service, 1985, Proposed land and resource management plan for the Medicine Bow National Forest and Thunder Basin National Grassland, Wyoming: U.S. Forest Service, Report no. 850523.

U.S. Forest Service, 1986, Cleveland National Forest land and resource management plan, Orange, Riverside, and San Diego Counties, California: U.S. Forest Service, Report no. 860209.

U.S. Forest Service, 1986, Croatan and Uwharrie National Forests land and resource management plan, North Carolina: U.S. Forest Service, Report no. 860218.

U.S. Forest Service, 1986, National Forests in Alabama Land and resource management plan: U.S. Forest Service, Report no. 860095.

U.S. Forest Service, 1986, Proposed Carson National Forest plan, Rio Arriba, Taos, Colfax, and Mora Counties, New Mexico: U.S. Forest Service, Report no. 860447.

U.S. Forest Service, 1986, Proposed Santa Fe National Forest land and resource management plan, Mora, San Miguel, Santa Fe, Sandoval, Los Alamos, and Rio Arriba Counties, New Mexico: U.S. Forest Service, Report no. 860013.

U.S. Forest Service, 1987, Angeles National Forest land and resource management plan, Los Angeles, Ventura, and San Bernardino Counties, California: U.S. Forest Service, Report no. 870394.

U.S. Forest Service, 1987, Environmental impact statement for the Apache-Sitgreaves National Forests Plan: U.S. Forest Service, Report, 392 p.

U.S. Forest Service, 1987, Santa Fe National Forest land management plan, Mora, San Miguel, Santa Fe, Sandoval, Los Alamos, and Rio Arriba Counties, New Mexico: U.S. Forest Service, Report no. 870299.

U.S. Forest Service, 1988, Proposed Kaibab National Forest land and resource management plan, Coconino, Yavapai, and Mohave Counties, Arizona: U.S. Forest Service, Report no. 880109.

U.S. Forest Service, 1988, Record of decision for USDA, Forest Service—Final environmental impact statement, Allegheny National Forest, land and resource management plan—Elk, Forest, McKean, and Warren Counties, Pennsylvania: U.S. Forest Service, Technical Report, 41 p.

U.S. Forest Service, 1989, San Bernardino National Forest land and resource management plan, San Bernardino and Riverside Counties, California: U.S. Forest Service, Report no. 890023.

U.S. Forest Service, 1990, Cherokee National Forest land and resource management plan, Tennessee, North Carolina, and Virginia (final supplement to the final environmental impact statement of April 1986): U.S. Forest Service, Report no. 900062, 62 p.

U.S. Forest Service, 1990, Umatilla National Forest land and resource management plan, Oregon and Washington: U.S. Forest Service, Report no. 900237.

U.S. Forest Service, 1991, Corral Mountain timber sale, San Juan National Forest, Archuleta County, Colorado: U.S. Forest Service, Report no. 910124, 59 p.

U.S. Forest Service, 1992, Martinez Creek timber sale, San Juan National Forest, Colorado: U.S. Forest Service, Report no. 920275, 75 p.

U.S. Forest Service, 1992, Norbeck Wildlife Preserve land management plan, Black Hills National Forest, Custer and Pennington Counties, South Dakota (final supplement to the environmental impact statement of July 1989): U.S. Forest Service, Report no. 920352, 92 p.

U.S. Forest Service, 1992, Rangeland ecosystem management in the Uinta National Forest, Provo, Utah: U.S. Forest Service, Report no. 920322, 189 p.

U.S. Forest Service, 1993, Six Rivers National Forest plan, Humboldt, Del Norte, Siskiyou, and Trinity Counties, California: U.S. Forest Service, Report no. 930340.

U.S. Forest Service, 1994, Oregon Dunes National Recreation Area management plan, Siuslaw National Forest—Coos, Douglas, and Lane Counties, Oregon: U.S. Forest Service, Report no. 940292.

U.S. Forest Service, 1994, Revised Land and resource management plan for the national forests and grasslands in Texas—Angelina, Fannin, Houston, Jasper, Montague,

Montgomery, Nacogdoches, Newton, Sabine, San Augustine, San Jacinto, Shelby, Trinity, Walker, and Wise Counties, Texas: U.S. Forest Service, Report no. 940385.

U.S. Forest Service, 1995, Draft environmental impact statement, Fish Bate analysis area—North Fork Ranger District, Clearwater National Forest, Clearwater County, Idaho: U.S. Forest Service, Technical Report.

U.S. Forest Service, 1995, Mendocino National Forest land and resource management plan—Colusa, Glenn, Lake, Mendocino, Tehama, and Trinity Counties, California: U.S. Forest Service, Report no. 950337.

U.S. Forest Service, 1995, Snowy Trail re-route, Ventura County, California: U.S. Forest Service, Report no. 950117, 109 p.

U.S. Forest Service, 1996, Snowy Trail re-route, Los Padres National Forest, Ventura County, California: U.S. Forest Service, Report no. 960524.

U.S. Forest Service, 2000, Silver Creek integrated resource project, Emmett Ranger District, Boise National Forest, Boise and Valley Counties, Idaho: U.S. Forest Service, Report no. 000241, 264 p.

U.S. Forest Service, 2001, Land and resource management plan, Uinta National Forest—Juab, Sanpete, Toole, Utah, and Wasatch Counties, Utah: U.S. Forest Service, Report no. 010142.

U.S. Forest Service, 2001, Starbucky restoration project, Red River Ranger District, Nez Perce National Forest, Idaho County, Idaho: U.S. Forest Service, Report no. 010274.

U.S. Forest Service, 2002, Final environmental impact statement Amendment to the Land and resource management plan—Management direction for acquired lands in southeastern Oklahoma, Ouachita National Forest, McCurtain County, Oklahoma: U.S. Forest Service, Technical Report.

U.S. Forest Service, 2002, Sixshooter project, Emmett Ranger District, Boise National Forest, Gem County, Idaho: U.S. Forest Service, Report no. 020238, 201 p.

U.S. Forest Service, 2002, Uncompahgre National Forest travel plan revision, Gunnison, Hinsdale, Mesa, Montrose, Ouray, San Juan, and San Miguel Counties, Colorado (final supplement to the final environmental impact statement of April 2000): U.S. Forest Service, Report no. 020131, 92 p.

U.S. Forest Service, 2002, West Gold project, Sandpoint Ranger District, Idaho Panhandle National Forests, Bonner County, Idaho: U.S. Forest Service, Report no. 020484.

U.S. Forest Service, 2002, Whiskey Campo resource management project, Mountain Home Ranger District, Boise National Forest, Elmore County, Idaho: U.S. Forest Service, Report no. 020299, 151 p.

U.S. Forest Service, 2003, Big Bend Ridge vegetation management project and timber sale, Ashton/Island Park Ranger District, Caribou—Targhee National Forest, Freemont County, Idaho: U.S. Forest Service, Report no. 030225.

U.S. Forest Service, 2003, Upper Bear timber sale, Council Ranger District, Payette National Forest, Adams County, Idaho: U.S. Forest Service, Report no. 030319, 478 p.

U.S. Forest Service, 2004, French Face ecosystem restoration, Ninemile Ranger District, Lolo National Forest, Montana: U.S. Forest Service, Report no. 040336, 184 p.

U.S. Forest Service, 2005, Dean Project Area, Bearlodge Ranger District, Black Hills National Forest, Crook County, Wyoming: U.S. Forest Service, Report no. 050212, 192 p.

U.S. Forest Service, 2005, Rocky Mountain Ranger District travel management plan, Rocky Mountain Ranger District, Lewis and Clark National Forest, Glacier, Pondera, Teton, and Lewis and Clark Counties, Montana: U.S. Forest Service, Report no. 050236, 386 p.

U.S. Forest Service, U.S. Bureau of Land Management, and U.S. Fish and Wildlife Service, 1997, George Washington National Forest land and resource management plan, Virginia and West Virginia (final supplement to the final environmental impact statement of January 1993): U.S. Forest Service, *U.S. Bureau of Land Management, and U.S. Fish and Wildlife Service,* Report no. 970046.

U.S. National Park Service, 1991, Big Cypress National Preserve, Collier, Monroe, and Dade Counties, Florida: *U.S. National Park Service,* Report no. 910386.

U.S. National Park Service, 1998, Redwood national and state parks general management plan, Del Norte and Humboldt Counties, California: *U.S. National Park Service,* Report no. 980290, 474 p.

U.S. National Park Service, 1999, Big Cypress National Preserve, recreational off-road vehicle management plan—Collier, Dade, and Monroe Counties, Florida (draft supplement to the environmental impact statement of October 1991): *U.S. National Park Service,* Report no. 990285, 171 p.

U.S. National Park Service, 2000, Recreational Off-road Vehicle management plan, Big Cypress National Preserve—Collier, Miami-Dade, and Monroe Counties, Florida (final supplement to the final environmental impact statement of October 1991): *U.S. National Park Service,* Report no. 000275, 619 p.

U.S. Soil Conservation Service, 1990, Erosion, Grand Traverse County, Michigan: *U.S. Soil Conservation Service,* Technical Report.

Vogt, G., 1979, Adverse effects of recreation on sand dunes—A problem for coastal zone management: *Coastal Zone Management Journal New York,* v. 6, no. 1, p. 3 7–68.

Warburton, M., Fisher, R., and Hathaway, S., 2004, Anza-Borrego Desert State Park amphibian survey, 2000-2002: *U.S. Geological Survey,* Report no. PB2005 105820, 44 p.

Watson, J.J., Kerley, G.I.H., and McLachlan, A., 1996, Human activity and potential impacts on dune breeding birds in the Alexandria Coastal Dunefield: *Landscape and Urban Planning,* v. 34, no. 3–4, p. 3 15–322.

Welch, R., Madden, M., and Doren, R.F., 1999, Mapping the Everglades: *Photogrammetric Engineering and Remote Sensing,* v. 65, no. 2, p. 163–170.

Wemple, B.C., Jones, J., and Grant, G., 1996, Channel network extension by logging roads in two basins, western Cascades, Oregon: *Water Resources Bulletin,* v. 32, no. 6, p. 1195–1207.

Wester, L., 1994, Weed management and the habitat protection of rare species—A case study of the endemic Hawaiian fern *Marsilea villosa*: *Biological Conservation,* v. 68, no. 1, p. 1–9.

Wheeler, A.P., Angermeier, P.L., and Rosenberger, A.E., 2005, Impacts of new highways and subsequent landscape urbanization on stream habitat and biota: *Reviews in Fisheries Science,* v. 13, no. 3, p. 141–164.

Wilson, J.P., and Seney, J.P., 1994, Erosional impact of hikers, horses, motorcycles, and off-road bicycles on mountain trails in Montana: *Mountain Research and Development,* v. 14, no. 1, p. 77–88.

Wolcott, T.G., and Wolcott, D.L., 1984, Impact of off-road vehicles on macroinvertebrates of a mid-Atlantic beach: *Biological Conservation,* v. 29, no. 3, p. 2 17–240.

Zamora, D., and Porter, M.D., 2000, Improved equipment for sampling dissolved oxygen at levee-pond commercial aquaculture facilities: *North American Journal of Aquaculture,* v. 62, no. 4, p. 311–315.

1.5. OHV Effects on Air Quality

Adams, E.S., 1975, Effects of lead and hydrocarbons from snowmobile exhaust on brook trout (*Salvelinus fontanlis*): *Transactions of the American Fisheries Society,* v. 104, no. 2, p. 363–373.

Agrawal, Y.K., Patel, M.P., and Merh, S.S., 1981, Lead in soils and plants—Its relationship to traffic volume and proximity to highway (Lalbag, Baroda City): *International Journal of Environmental Studies,* v. 16, no. 3–4, p. 222–224.

Al-Awadhi, J.I., 2001, Impact of gravel quarrying on the desert environment of Kuwait: *Environmental Geology,* v. 41, no. 3–4, p. 365–371.

Andrews, A., 1990, Fragmentation of habitat by roads and utility corridors—A review: *Australian Zoologist,* v. 26, p. 130–141.

Angold, P.G., 1997, The impact of a road upon adjacent heathland vegetation—Effects on plants species composition: *Journal of Applied Ecology,* v. 34, no. 2, p. 409–417.

Bazzaz, F.A., and Garbutt, K., 1988, The response of annuals in competitive neighborhoods— Effects of elevated CO2: *Ecology,* v. 69, no. 4, p. 937–946.

Benson, P.E., Nokes, W.A., Cramer, R.L., O`Connor, J., Lindeman, W., 1986, Air-quality and noise issues in environmental planning: Washington, D.C., T*ransportation Research Board and National Research Council Technical Report,* 72 p., *http://www.osti.gov/energycitations/product.biblio.jsp?osti_id=7172765.*

Berry, K.H., 1980, The effects of four-wheel vehicles on biological resources—Soils, vegetation, small animals, management problems, Southwestern United States, *in* Andrews, R.N.L., and Nowak, P.F., eds., Off-road vehicle use—A management challenge: Washington D.C., *U.S. Office of Environmental Quality,* p. 23 1–233.

Bishop, G.A., Morris, J.A., and Stedman, D.H., 2001, Snowmobile contributions to mobile source emissions in Yellowstone National Park: *Environmental Science and Technology,* v. 35, no. 14, p. 2874–2881.

Boer, B., 1998, Anthropogenic factors and their potential impacts on the sustainable development of Abu Dhabi's terrestrial biological resources: *International Journal of Sustainable Development and World Ecology,* v. 5, no. 2, p. 125–135.

Brown, G., and Porembski, S., 2000, Phytogenic hillocks and blow-outs as 'safe sites' for plants in an oil-contaminated area of northern Kuwait: *Environmental Conservation,* v. 27, no. 3, p. 242–249.

Byrd, D.S., Gilmore, J.T., and Lea, R.H., 1983, Effect of decreased use of lead in gasoline on the soil of a highway: *Environmental Science and Technology,* v. 17, p. 12 1–123.

Cadle, S.H., Belian, T.C., Black, K.N., Minassian, F., Natarajan, M., Tierney, E.J., and Lawson, D.R., 2005, Real-world vehicle emissions—A summary of the 14th Coordinating Research Council on Road Vehicle Emissions Workshop: *Journal of the Air and Waste Management Association,* v. 55, no. 2, p. 130–146.

Cadle, S.H., Croes, B.E., Minassian, F., Natarajan, M., Tierney, E.J., and Lawson, D.R., 2004, Real-world vehicle emissions—A summary of the 13th Coordinating Research Council on Road Vehicle Emissions Workshop: *Journal of the Air and Waste Management Association* v. 54, no. 1, p. 8–23.

Dennis, D.F., 1998, Analyzing public inputs to multiple objective decisions on national forests using conjoint analysis: *Forest Science,* v. 44, no. 3, p. 421–428.

Dolfay, M.F., 1980, Methods of improving the management of four-wheel vehicles Wenatchee National Forest, *in* Andrews, R.N.L., and Nowak, P.F., eds., Off-road vehicle use—A management challenge: Washington, D.C., U.S. *Office of Environmental Quality,* p. 259–261.

Dunnell, C.W., 1980, Protecting and rehabilitating ORV off-road recreational vehicles use areas—Erosion control and management, Wenatchee National Forest, *in* Andrews, R.N.L., and Nowak, P.F., eds., Off-road vehicle use—A management challenge: Washington, D.C., *U.S. Office of Environmental Quality,* p. 100–102.

Falkengren-Grerup, U., 1986, Soil acidification and its impact on ground vegetation: *Oecologia,* v. 70, p. 339–347.

Ferris, R., and Taylor, G., 1995, Contrasting effects of elevated CO^2 and water deficit on two native herbs: *New Phytologist,* v. 131, no. 4, p. 491–501.

Fox, D.G., 1986, Establishing a baseline/protocols for measuring air quality effects in wilderness: *U.S. Forest Service,* Rocky Mountain Forest and Range Experimental Station, Technical Report, 85–9 1 p.

Gish, C.D., and Christensen, R.E., 1973, Cadmium, nickel, lead, and zinc in earthworms from roadside soil: *Environmental Science and Technology,* v. 7, p. 1060–1062.

Grue, C.E., O'Shea, T.J., and Hoffman, D.J., 1984, Lead concentrations and reproduction in highway-nesting barn swallows [*Hirundo rustica*]: *Condor,* v. 86, no. 4, p. 383–389.

Harrison, R.T., 1980, Environmental impact of off-road motorcycles, *in* Andrews, R.N.L., and Nowak, P.F., eds., Off-road vehicle use—A management challenge: Washington, D.C., *U.S. Office of Environmental Quality,* p. 266–269.

Holzapfel, C., and Schmidt, W., 1990, Roadside vegetation along transects in the Judean Desert: *Israel Journal of Botany,* v. 39, p. 263–270.

Hunt, A., Dickens, H.J., and Whelan, R.J., 1987, Movement of mammals through tunnels under railway lines: *Australian Zoologist,* v. 24, no. 2, p. 89–93.

Kay, J., 1981, Evaluating environmental impacts of off-road vehicles: *Journal of Geography,* v. 80, no. 1, p. 10–18.

Kean, A.J., Sawyer, R.F., and Harley, R.A., 2000, A fuel-based assessment of off-road diesel engine emissions (1995): *Journal of the Air and Waste Management Association,* v. 50, no. 11, p. 1929–1939.

Khalaf, F.I., 1989, Desertification and aeolian processes in the Kuwait Desert: *Journal of Arid Environments,* v. 16, no. 2, p. 125–145.

Larney, F.J., Leys, J.F., Mueller, J.F., and McTainsh, G.H., 1999, Dust and endosulfan deposition in a cotton-growing area of northern New South Wales, Australia: *Journal of Environmental Quality,* v. 28, no. 2, p. 692–701.

Lovich, J.E., and Bainbridge, D., 1999, Anthropogenic degradation of the southern California desert ecosystem and prospects for natural recovery and restoration: *Environmental Management,* v. 24, no. 3, p. 309–326.

Mitchell, G., 2005, Forecasting environmental equity—Air quality responses to road user charging in Leeds, UK.: *Journal of Environmental Management,* v. 77, no. 3, p. 2 12–226.

Mooney, H.A., Kueppers, M., Koch, G., Gorham, J., Chu, C., and Winner, W.E., 1988, Compensating effects to growth of carbon partitioning changes in response to SO2-induced photosynthetic reduction in radish: *Oecologia,* v. 75, no. 4, p. 502–506.

Motto, H.L., Daines, R.H., Chilko, D.M., and Motto, C.K., 1970, Lead in soils and plants—Its relationship to traffic volume and proximity to highways: *Environmental Science and Technology,* v. 4, p. 23 1–237.

Nash, A.E.K., 1980, Nature aesthetics, the public interest, and ORV off-road recreational vehicles users' perspectives, *in* Andrews, R.N.L., and Nowak, P.F., eds., Off-road vehicle use—A management challenge: Washington, D.C., *U.S. Office of Environmental Quality,* p. 12–77.

Nicholes, G.E., 1980, A foundation for problem solutions—Off-road recreational vehicles, user attitudes, management problems, *in* Andrews, R.N.L., and Nowak, P.F., eds., Off-road vehicle use—A management challenge: Washington, D.C., *U.S. Office of Environmental Quality,* p. 103–109.

Payne, G.F., Leininger, W.D., and Foster, J., 1979, *How off-road vehicles affect range quality. When driving across rangelands, it's better to blaze a new trail: Bozeman, Montana,* Montana Agricultural Experiment Station, Technical Report, 2 p.

Quarles, H.D., Hanawalt, R.B., and Odum, W.E., 1974, Lead in small mammals, plants, and soil at varying distances from a highway: *Journal of Applied Ecology,* v. 11, no. 3, p. 937–949.

Samaras, Z., and Zierock, K.H., 1995, Off-road vehicles—A comparison of emissions with those from road transport: *Science of the Total Environment,* v. 169, p. 249–253.

Schade, G.A.J., 1980, Four-wheel vehicle user's perspective, *in* Andrews, R.N.L., and Nowak, P.F., eds., *Off-road vehicle use—A management challenge:* Washington, D.C., U.S. Office of Environmental Quality.

Smith, W.H., 1976, Lead contamination of the roadside ecosystem: *Journal of the Air Pollution Control Association,* v. 26, p. 753–766.

Spencer, H.J., and Port, G.R., 1988, Effects of roadside conditions on plants and insects: II. Soil conditions: *Journal of Applied Ecology,* v. 25, no. 22, p. 709–7 15.

Thompson, C.R., Olszyk, D.M., Kats, G., Bytnerowicz, A., Dawson, P.J., and Wolf, J.W., 1984, Effects of ozone or sulfur dioxide on annual plants of the Mojave Desert: *Journal of the Air Pollution Control Association,* v. 34, no. 10, p. 1017–1022.

Tocher, S.R., 1980, Four-wheel vehicles—Summary management on public lands, *in* Andrews, R.N.L., and Nowak, P.F., eds., Off-road vehicle use—A management challenge: Washington, D.C., *U.S. Office of Environmental Quality,* p. 286–289.

U.S. Army Corps of Engineers, 1994, Stationing of mechanized or armored combat forces at Fort Lewis, Thurston and Pierce Counties, Washington: *U.S. Army Corps of Engineers,* Report no. 940045.

U.S. Army Corps of Engineers, 2001, Fort Bliss mission and master plan—Dona Ana and Otero Counties, New Mexico, and El Paso County, Texas: *U.S. Army Corps of Engineers,* Report no. 010081.

U.S. Bureau of Land Management, 1985, Grand Junction resource area—Resource management plan and environmental impact statement: *U.S. Bureau of Land Management*, Report, 171 p.

U.S. Bureau of Land Management, 1986, Baker resource management plan, Baker County, Oregon: *U.S. Bureau of Land Management,* Report no. 860150, 129 p.

U.S. Bureau of Land Management, 1986, Lander resource management plan, Lander, Wyoming: *U.S. Bureau of Land Management,* Report no. 860449.

U.S. Bureau of Land Management, 1986, Proposed Baker resource management plan, Vale District, Oregon: *U.S. Bureau of Land Management,*, Report no. 860396.

U.S. Bureau of Land Management, 1986, Utah statewide wilderness environmental impact statement: *U.S. Bureau of Land Management,*, Report no. 860036.

U.S. Bureau of Land Management, 1987, North Dakota resource management plan, Dunn and Bowman Counties, North Dakota: *U.S. Bureau of Land Management,*, Report no. 870236, 199 p.

U.S. Bureau of Land Management, 1987, Pinedale resource area resource management plan, Sublette and Lincoln Counties, Wyoming: *U.S. Bureau of Land Management,*, Report no. 870077, 347 p.

U.S. Bureau of Land Management, 1987, Pinedale resource area resource management plan, Sublette, Teton, Lincoln, and Fremont Counties, Wyoming: *U.S. Bureau of Land Management,*, Report no. 870428, 232 p.

U.S. Bureau of Land Management, 1988, Challis resource area proposed resource management plan, upper Columbia-Salmon Clearwater Districts, Custer and Lemhi Counties, Idaho: *U.S. Bureau of Land Management,*, Report no. 980508.

U.S. Bureau of Land Management, 1988, Cody resource area resource management plan, Big Horn and Park Counties, Wyoming: *U.S. Bureau of Land Management,*, Report no. 880315.

U.S. Bureau of Land Management, 1988, Pony Express resource management plan, Toole, Utah, and Salt Lake Counties, Utah: *U.S. Bureau of Land Management,*, Report no. 880310, 147 p.

U.S. Bureau of Land Management, 1988, Proposed Las Vegas resource management plan and final environmental impact statement, Clark and Nye Counties, Nevada: *U.S. Bureau of Land Management,*, Report no. 980207.

U.S. Bureau of Land Management, 1988, Uncompahgre Basin resource management plan, Colorado: *U.S. Bureau of Land Management,*, Report no. 880345.

U.S. Bureau of Land Management, 1988, Wilderness study areas in the Rock Springs District, Fremont, Lincoln, Sublette, and Sweetwater Counties, Wyoming (revised Draft environmental impact statement): *U.S. Bureau of Land Management,*, Report no. 880359, 175 p.

U.S. Bureau of Land Management, 1989, Dixie resource area management plan, Washington County, Utah: *U.S. Bureau of Land Management,*, Report no. 890290, 250 p.

U.S. Bureau of Land Management, 1989, San Rafael resource management plan, Emery County, Utah: *U.S. Bureau of Land Management,*, Report no. 890240.

U.S. Bureau of Land Management, 1989, White Sands resource management plan Amendment for McGregor Range, Otero County, New Mexico: *U.S. Bureau of Land Management,*, Report no. 890127.

U.S. Bureau of Land Management, 1990, Rock Springs District wilderness study areas, Fremont, Lincoln, Sublette, and Sweetwater Counties, Wyoming: *U.S. Bureau of Land Management,*, Report no. 900439, 324 p.

U.S. Bureau of Land Management, 1991, Safford District resource management plan, Arizona: *U.S. Bureau of Land Management,*, Report no. 910314, 511 p.

U.S. Bureau of Land Management, 1994, Stateline resource management plan, Clark and Nye Counties, Nevada (draft environmental impact statement and draft supplement): *U.S. Bureau of Land Management,*, Report no. 940177.

U.S. Bureau of Land Management, 1998, Socorro resource area resource management plan, New Mexico: *U.S. Bureau of Land Management,*, Report no. 880397, 266 p.

U.S. Bureau of Land Management, 2000, Pinedale anticline oil and gas exploration and development project, Sublette County, Wyoming: *U.S. Bureau of Land Management,*, Report no. 000159, 301 p.

U.S. Bureau of Land Management, 2001, National management strategy for motorized off-highway vehicle use on public lands: *U.S. Bureau of Land Management,*, Report no. PB2001103 162, 58 p.

U.S. Bureau of Land Management, 2004, Dillon resource management plan, Beaverhead and Madison Counties, Montana: *U.S. Bureau of Land Management,*, Report no. 040153, 626 p.

U.S. Bureau of Land Management, 2004, Jack Morrow Hills coordinated activity plan, Sweetwater, Fremont, and Sublette Counties, Wyoming: *U.S. Bureau of Land Management,*, Technical Report.

U.S. Bureau of Land Management, 2005, Newmont Mining Company, Emigrant project, Elko County, Nevada: *U.S. Bureau of Land Management,*, Report no. 050117, 269 p.

U.S. Bureau of Reclamation, 2001, Pothole Reservoir resource management plan, Grant County, Washington: U.S. Bureau of Reclamation, Report no. 010518, 567 p.

U.S. Fish and Wildlife Service, 1986, Great Dismal Swamp National Wildlife Refuge master plan, Virginia and North Carolina: U.S. Fish and Wildlife Service, Report no. 860501, 245 p.

U.S. Forest Service, 1973, Proposed off-road vehicle regulations and administrative instructions: U.S. Forest Service, Report no. USDAFS–FES(ADM)–73–49, 110 p.

U.S. Forest Service, 1973, Proposed regulations and administrative instructions relating to use of off-road vehicles on national forest lands: U.S. Forest Service, Report no. USDAFS–DES(ADM)–73–49, 22 p.

U.S. Forest Service, 1985, Kisatchie National Forest land and resource management plan, Louisiana: U.S. Forest Service, Report no. 850414.

U.S. Forest Service, 1985, Proposed land and resource management plan for the Medicine Bow National Forest and Thunder Basin National Grassland, Wyoming: U.S. Forest Service, Report no. 850523.

U.S. Forest Service, 1986, Land and resource management plan, Chippewa National Forest, Beltrami, Cass, and Itasca Counties, Minnesota: U.S. Forest Service, Report no. 860208.

U.S. Forest Service, 1986, Proposed Santa Fe National Forest land and resource management plan, Mora, San Miguel, Santa Fe, Sandoval, Los Alamos, and Rio Arriba Counties, New Mexico: U.S. Forest Service, Report no. 860013.

U.S. Forest Service, 1987, Proposed Apache-Sitgreaves National Forests land and resource management plan, Apache, Coconino, Greenlee, and Navajo Counties, Arizona: U.S. Forest Service, Report no. 870387.

U.S. Forest Service, 1987, Proposed land and resource management plan for the Coconino Forest, Coconino, Gila, and Yavapai Counties, Arizona: U.S. Forest Service, Report no. 870288.

U.S. Forest Service, 1987, Santa Fe National Forest land management plan, Mora, San Miguel, Santa Fe, Sandoval, Los Alamos, and Rio Arriba Counties, New Mexico: U.S. Forest Service, Report no. 870299.

U.S. Forest Service, 1989, San Bernardino National Forest land and resource management plan, San Bernardino and Riverside Counties, California: U.S. Forest Service, Report no. 890023.

U.S. Forest Service, 1993, Six Rivers National Forest plan, Humboldt, Del Norte, Siskiyou, and Trinity Counties, California: U.S. Forest Service, Report no. 930340.

U.S. Forest Service, 1994, Revised land and resource management plan for the national forests and grasslands in Texas—Angelina, Fannin, Houston, Jasper, Montague, Montgomery, Nacogdoches, Newton, Sabine, San Augustine, San Jacinto, Shelby, Trinity, Walker, and Wise Counties, Texas: U.S. Forest Service, Report no. 940385.

U.S. Forest Service, 1995, Mendocino National Forest land and resource management plan—Colusa, Glenn, Lake, Mendocino, Tehama, and Trinity Counties, California: U.S. Forest Service, Report no. 950337.

U.S. Forest Service, 1995, Snowy Trail re-route, Ventura County, California: U.S. Forest Service, Report no. 950117, 109 p.

U.S. Forest Service, 1996, Snowy Trail re-route, Los Padres National Forest, Ventura County, California: U.S. Forest Service, Report no. 960524.

U.S. Forest Service, 2001, Land and resource management plan, Uinta National Forest—Juab, Sanpete, Toole, Utah, and Wasatch Counties, Utah: U.S. Forest Service, Report no. 010142.

U.S. Forest Service, 2001, Starbucky restoration project, Red River Ranger District, Nez Perce National Forest, Idaho County, Idaho: U.S. Forest Service, Report no. 010274.

U.S. Forest Service, 2002, Sixshooter project, Emmett Ranger District, Boise National Forest, Gem County, Idaho: U.S. Forest Service, Report no. 020238, 201 p.

U.S. Forest Service, 2002, West Gold project, Sandpoint Ranger District, Idaho Panhandle National Forests, Bonner County, Idaho: U.S. Forest Service, Report no. 020484.

U.S. Forest Service, 2002, Whiskey Campo resource management project, Mountain Home Ranger District, Boise National Forest, Elmore County, Idaho: U.S. Forest Service, Report no. 020299, 151 p.

U.S. Forest Service, 2003, Big Bend Ridge vegetation management project and timber sale, Ashton/Island Park Ranger District, Caribou-Targhee National Forest, Freemont County, Idaho: U.S. Forest Service, Report no. 030225.

U.S. Forest Service, 2003, Upper Bear timber sale, Council Ranger District, Payette National Forest, Adams County, Idaho: U.S. Forest Service, Report no. 030319, 478 p.

U.S. Forest Service, 2004, French Face ecosystem restoration, Ninemile Ranger District, Lolo National Forest, Montana: U.S. Forest Service, Report no. 040336, 184 p.

U.S. Forest Service, 2005, Dean project area, Bearlodge Ranger District, Black Hills National Forest, Crook County, Wyoming: U.S. Forest Service, Report no. 050212, 192 p.

U.S. Forest Service, 2005, Woodrock project, Tongue Ranger District, Bighorn National Forest, Sheridan County, Wyoming: U.S. Forest Service, Report no. 050112, 201 p.

U.S. Forest Service, 2006, Hoosier National Forest land and resource management plan—Final environmental impact statement: U.S. Forest Service, Technical Report.

U.S. Naval Facilities Engineering Command, 2003, Advanced amphibious assault vehicle, Marine Corps Base Camp Pendleton and San Clemente Range Complex, San Diego County, California: U.S. Naval Facilities Engineering Command, Report no. 030201.

Udevitz, M.S., Howard, C.A., Robel, R.J., and Curnutte, B., 1980, Lead contamination in insects and birds near an interstate highway, Kansas: Environmental Entomology, v. 9, no. 1, p. 35–36.

Walker, D.A., and Everett, K.R., 1987, Road dust and its environmental impact on Alaskan taiga and tundra: Arctic and Alpine Research, v. 19, no. 4, p. 479–489.

Westec Services Inc., 1979, Fugitive dust impacts during off-road vehicle (ORV) events in the California desert: Tustin, California, WESTEC Services, Inc., Technical Report, 40 p.

Winner, W.E., and Atkinson, C.J., 1986, Absorption of air pollutants by plants and consequences: Trends in Ecology and Evolution, v. 1, p. 15–18.

1.6. Socioeconomic Implications of OHV Use

Absher, J.D., and Bright, A.D., 2004, Communication research in outdoor recreation and natural resources management, *in* Manfredo, M.J., Vaske, J.J., Bruyere, B.L., Field, D.R., and Brown, P.J., eds., Society and Natural Resources—A summary of knowledge: Jefferson, Missouri, Modern Litho, p.117–126.

Achana, F., 2005, 2005 ATV/motorbike user survey: Idaho Department of Parks and Recreation, Technical Report, 37 p.

Alaska Pulp Corporation, U.S. Forest Service, and U.S. Army Corps of Engineers, 1994, Ushk Bay timber sale, Alaska Pulp Corporation long-term timber sale Contract, Chatnam Area, Tongass National Forest, Alaska: U.S. Forest Service and U.S. Army Corps of Engineers, Report no. 940408.

American Motorcyclist Association, 1979, The economic impact of off-highway motorcycling in southern California and Mexico: Westerville, Ohio, Department of Government Relations, Technical Report, 14 p.

Bannister, E.N., 1979, Impact of road networks on southeastern Michigan lakeshore drainage: Water Resources Research, v. 15, no. 6, p. 1515–1520.

Bight, A.D., Coredell, H.K., Hoover, A.P., and Tarranr, M.A., 2003, A human dimensions framework—Guidelines for conducting social assessments: Asheville, North Carolina, U.S. Forest Service, Southern Research Station, 83 p.

Bjornlie, D.D., and Garrott, R.A., 2001, Effects of winter road grooming on bison in Yellowstone National Park: Journal of Wildlife Management, v. 65, no. 3, p. 560–572.

Boyle, S.A., and Samson, F.B., 1985, Effects of nonconsumptive recreation on wildlife—A review: Wildlife Society Bulletin, v. 13, p. 110–116.

Brooks, J.J., and Champ, P.A., 2006, Understanding the wicked nature of "unmanaged recreation" in Colorado's Front Range: Environmental Management, v. 38, no. 5, p. 784–798, *http://www.springerlink.com/content/ek62272u626u6074/fulltext.pdf.*

Brown, A.C., and McLachlan, A., 2002, Sandy shore ecosystems and the threats facing them— Some predictions for the year 2025: Environmental Conservation, v. 29, no. 1, p. 62–77.

Buckley, R., 2004, Environmental impacts of ecotourism: Wallingford, England, CABI Publishing, 389 p.

Buckley, R., 2004, Environmental impacts of motorized off-highway vehicles, *in* Buckley, R., ed., Environmental impacts of ecotourism: Wallingford, England, CABI Publishing, p. 83–97.

Buckley, R., and King, N., 2003, Visitor-impact data in a land-management context, *in* Buckley, R.C., Pickering, D., and Weaver, D.B., eds., Nature-based tourism, environment and land management: Wallingford, England, CABI Publishing, p. 89-99.

Burger, J., 1986, The effect of human activity on shorebirds in two coastal bays in northeastern United States: Environmental Conservation, v. 13, no. 2, p. 123–130.

Bury, R.L., and Fillmore, E.R., 1974, Design of motorcycle areas near campgrounds—Effects on riders and non riders: College Station, Texas, Department of Recreation and Parks, Texas A & M University, Technical Report, 72 p.

Bury, R.L., Holland, S.M., and McEwen, D.N., 1983, Analyzing recreational conflict—Understanding why conflict occurs is requisite to managing that conflict: Journal of Soil and Water Conservation, v. 3, no. 5, p. 401–403.

Bury, R.L., Wendling, R.C., and McCool, S.F., 1976, Off-road recreation vehicles—A research summary, 1969–1975: College Station, Texas, Texas Agricultural Experiment Station, Technical Report.

Call, C.A., Barker, J.R., and McKell, C.M., 1981; Visitor impact assessment of scenic view areas at Bryce Canyon National Park: Journal of Soil and Water Conservation, v. 36, no. 1, p. 50–53.

Cole, D.N., 2004, Environmental impacts of outdoor recreation in wildlands, *in* Manfredo, M.J., Vaske, J.J., Bruyere, B.L., Field, D.R., and Brown, P.J., eds., Society and natural resources—A summary of knowledge: Jefferson, Missouri, Modern Litho, p. 107–116.

Cole, D.N. and Landres, P.B., 1995, Indirect effects of recreation on wildlife, *in* Knight, R.L., and Gutzwiller, K.J., eds., Wildlife and recreationists—Coexistence through management and research: Washington, D.C., Island Press, p. 183–202.

Collins, E., O'Farrell, T.P., and Rhoads, W., 1982, Annotated bibliography for biologic overview for the Nevada nuclear waste storage investigations, Nevada Test Site, Nye County, Nevada: Goleta, California, EG and G, Inc., Report no. EGG1 1832419, 48 p.

Collins, E., O'Farrell, T.P., and Rhoads, W., 1982, Biologic overview for the Nevada Nuclear Waste Storage investigations, Nevada Test Site, Nye County, Nevada: Goleta, California, EG and G, Inc., Report no. EGG1 1832460, 55 p.

Cordell, H.K., Betz, C.J., Green, G., and Owens, M., 2005, Off-highway vehicle recreation in the United States, regions, and states—A national report from the national survey on recreation and the environment (NSRE): U.S. Forest Service, Southern Research Station, Technical Report, 90 p.

Council on Environmental Quality, 1997, Considering cumulative effects under the National Environmental Policy Act: Washington, D.C., Council on Environmental Quality, Technical Report, *http://www.nepa.gov/nepa/ccenepa/ccenepa.htm.*

Crimmins, T., 1999, Colorado off-highway vehicle user survey—Summary of results: Denver, Colorado, Colorado State Parks, Technical Report.

Dave Miller Associates, 1981, An economic/social assessment of snowmobiling in Maine: Windham, Maine, Dave Miller Associates, Technical Report, 52 p.

Dawson, K.J., 1985, Natural area planning for recreational use transition: Landscape Planning, v. 12, no. 2, p. 111–123.

Dean Runyan Associates, 2000, Campers in California travel patterns and economic impacts: Portland, Oregon, Dean Runyan Associates, Technical Report, 76 p.

Duever, M.J., Riopelle, L.A., and McCollom, J.M., 1986, Long term recovery of experimental off-road vehicle impacts and abandoned old trails in the Big Cypress National Preserve: Naples, Florida, National Audubon Society, Technical Report, 56 p.

Ewert, A., and Shultis, J., 1999, Technology and backcountry recreation—Boon to recreation or bust for management?: Technology and Leisure, v. 70, no. 8, p. 23–31.

Fisher, A.L., Blahna, D.J., and Bahr, R., 2001, Off-highway vehicle uses and owner preferences in Utah: Logan, Utah, Institute for Outdoor Recreation and Tourism, Department of Forest Resources, Utah State University, Report no. IORT PR2001–02, 80 p.

Graefe, A.R., and Thapa, B., 2004, Conflict in natural resource recreation, *in* Manfredo, M.J., Vaske, J.J., Bruyere, B.L., Field, D.R., and Brown, P.J., eds., Society and natural resources—A summary of knowledge: Jefferson, Missouri, Modern Litho, p. 209–224.

Gray, J.R., 1977, Kinds and costs of recreational pollution in the Sandia Mountains: New Mexico Agricultural Experiment Station Bulletin, Technical Report, 57 p.

Gutzwiller, K.J., 1995, Recreational disturbances and wildlife communities *in* Knight, R.L., and Gutzwiller, K.J., eds., Wildlife and recreationists—Coexistence through management and research: Washington, D.C., Island Press, p. 169–181.

Hall, C., and Dearden, P., eds., 1984, The impact of "non-consumptive" recreation on wildlife— An annotated bibliography: Monticello, Illinois, Vance Bibliographies, 45 p.

Heede, B.H., 1983, Control of rills and gullies in off-road vehicle traffic areas, *in* Webb, R.H., and Wilshire, H.G., eds., Environmental effects of off-road vehicles—Impacts and management in arid regions: New York, Springer-Verlag, p. 245–264.

Hendon, W.C., 1991, The wilderness as a source of recreation and renewal—Who uses it? what are their characteristics? their other interests? their preferences?: American Journal of Economics and Sociology, v. 50, no. 1, p. 105–112.

Hosier, P.E., and Eaton, T.E., 1980, The impact of vehicles on dune and grassland vegetation on a southeastern North Carolina barrier beach: Wilmington, North Carolina, North Carolina University at Wilmington, Report no. UNC–SG–R–1 64, 13 p.

Hosier, P.E., Kochlar, M., and Thayer, V., 1981, Off-road vehicle and pedestrian track effects on the sea-approach of hatchling loggerhead turtles: Environmental Conservation, v. 8, no. 2, p. 158–161.

Iverson, R.M., Hinckley, B.S., and Webb, R.M., 1981, Physical effects of vehicular disturbances on arid landscapes: Science, v. 212, no. 4497, p. 915–917.

Jim, C., 1989, Visitor management in recreation areas: Environmental Conservation, v. 16, no. 1, p. 19–32.

Jim, C.Y., 1987, Trampling impacts of recreationists on picnic sites in a Hong Kong county park: Environmental Conservation, v. 14, no. 2, p. 117–127.

Kant, H., and Sood, K.G., 1974, A study on off-road transportation by tractors: Indian Forester, v. 100, no. 5, p. 327–341.

Karanja, G., 2003, Tourism impacts in Masai Mara National Reserve: IIED Wildlife and Development Series, v. 14, p. 5–16.

Kato, T.T., and O'Farrell, T.P., 1986, Biological assessment of the effects of petroleum production at maximum efficient rate, Naval Petroleum Reserve No. 1 (Elk Hills), Kern County, California, on the endangered blunt-nosed leopard lizard, *Gambelia silus*: Goleta, California, EG and G Energy Measurements, Inc., Report no. EGG102 822 108, 72 p.

Kay, J., 1981, Evaluating environmental impacts of off-road vehicles: Journal of Geography, v. 80, no. 1, p. 10–18.

Kirkpatrick, J.B., 2001, Ecotourism, local and indigenous people, and the conservation of the Tasmanian Wilderness World Heritage Area: Journal of the Royal Society of New Zealand, v. 31, no. 4, p. 819–829.

Knight, R.L., and Cole, D.N., 1995, Wildlife response to recreationists, *in* Knight, R.L., and Gutzwiller, K.J., eds., Wildlife and recreationists—Coexistence through management and research: Washington, D.C., Island Press, p. 5 1–69.

Koeppel, W., and Strauss, C., 1984, Areal evaluation of vehicle mobility: Frankfurt, Germany, Battelle Institute, Technical Report, 20 p.

Landsberg, J., Logan, B., and Shorthouse, D., 2001, Horse riding in urban conservation areas— Reviewing scientific evidence to guide management: Ecological Management and Restoration, v. 2, no. 1, p. 36–46.

Lewis, O., 1969, Arid lands and their future, *in* Bender, G.L., ed., Future environments of arid regions of the southwest, contribution no. 12, American Association for the Advancement of Science Committee on Desert and Arid Zone Research symposium, Colorado Springs, Colorado, May 7-10, 1969: Washington, D.C., American Association for the Advancement of Science, Committee on Desert and Arid Zone Research, p. 33–3 8.

Losos, E., Hayes, J., Phillips, A., Alkire, C., and Wilcove, D., 1993, Taxpayers double burden— Federal resource subsidies and endangered species: Missoula, Montana, The Wilderness Society's Bolle Center for Ecosystem Management, Technical Report, 95 p.

Lovich, J.E., and Bainbridge, D., 1999, Anthropogenic degradation of the southern California desert ecosystem and prospects for natural recovery and restoration: Environmental Management, v. 24, no. 3, p. 309–326.

Lucas, R.C., 1978, Impact of human pressure on parks, wilderness, and other recreation lands, Sourcebook on the environment—A guide to the literature: Chicago, University of Chicago Press, p. 221–239.

Luckenbach, R.A., 1978, An analysis of off-road vehicle use on desert avifaunas: Transactions of the North American Wildlife and Natural Resources Conference, v. 43, p. 157–162.

Major, M.J., 1987, Managing off-the-road vehicles—A recurring round of events: Journal of Forestry, v. 85, no. 11, p. 37-41.

Matchett, J.R., Gass, L., Brooks, M.L., Mathie, A.M., Vitales, R.D., Campagna, M.W., Miller, D.M., and Weigand, J.F., 2004, Spatial and temporal patterns of off-highway vehicle use at the Dove Springs OHV open area, California: U.S. Geological Survey, Report, 17 p.

Meyer, K.G., 2002, Managing degraded off-highway vehicle trails in wet, unstable, and sensitive environments: U.S. Department of Interior, National Park Service, Report no. PB2005 105502, 88 p.

Meyer, K.G., 2002, Proposed best management practices for OHV/ATV trail management in Alaska: U.S. National Park Service Rivers Trails and Conservation Assistance Program, Technical Report, 19 p.

Mitchell, G., 2005, Forecasting environmental equity—Air quality responses to road user charging in Leeds, UK.: Journal of Environmental Management, v. 77, no. 3, p. 2 12–226.

Moorhead, D.L., Linkins, A.E., and Everett, K.R., 1996, Road dust alters extracellular enzyme activities in tussock tundra soils, Alaska, U.S.A.: Arctic and Alpine Research, v. 28, no. 3, p. 346-351.

Mortensen, C.O., 1989, Visitor use impacts within the Knobstone Trail corridor: Journal of Soil and Water Conservation, v. 44, no. 2, p. 156–159.

Muleski, G.E., Garmen, G., and Cowherd, C., 1994, Surface coal mine emission factor field study: Kansas City, Missouri, Midwest Research Institute, Report no. EPA/454/R95/010, 167 p.

Nelson, C.M., and Lynch, J.A., 2001, A usable pilot off-road vehicle project evaluation: East Lansing, Michigan, Department of Park, Recreation and Tourism Resources, Michigan State University, Technical Report, 50 p.

Nelson, C.M., Lynch, J.A., and Stynes, D.J., 2000, Michigan licensed off-road vehicle use and users—1998–99: East Lansing, Michigan, Department of Park, Recreation and Tourism Resources, Michigan State University, Technical Report, 49 p.

Noake, D.W., 1967, Camping as a factor in the ecological impact of tourism and recreation: International Union Conservation, p. 224–229.

Noe, F.P., Hammitt, W.E., and Bixler, R.D., 1995, Park user perceptions of resource and use impacts under varied situations in three National Parks: Journal of Environmental Management, v. 49, no. 3, p. 323–336.

Nova Scotia Voluntary Planning Task Force on Off-Highway Vehicles, 2003, Off-highway vehicles in Nova Scotia: Nova Scotia Voluntary Planning Task Force on Off-Highway Vehicles, Technical Report, 12 p.

O'Farrell, T.P., 1984, Conservation of the endangered San Joaquin kit fox *Vulpes macrotis mutica* in the Naval Petroleum Reserves, California: Acta Zoologica Fennica, v. 172, p. 207–208.

O'Farrell, T.P., Harris, C.E., Kato, T.T., and McCue, P.M., 1986, Biological assessment of the effects of petroleum production at maximum efficient rate, Naval Petroleum Reserve No. 1 (Elk Hills), Kern County, California, on the endangered San Joaquin kit fox, *Vulpes macrotis mutica*: Goleta, California, EG and G Energy Measurements, Inc., Report no. EGG102822107, 84 p.

O'Farrell, T.P., and Kato, T.T., 1987, Biological assessment of the effects of petroleum production activities, Naval Petroleum Reserves in California, on the endangered giant kangaroo rat, *Dipodomys ingens*: Goleta, California, EG and G Energy Measurements, Inc., Report no. EGG102822183, 34 p.

Pelton, C., 1987, A computer program for hill-shading digital topographic data sets: Computers and Geosciences, v. 13, no. 5, p. 545–548.

Propst, D.B., Shomaker, J.H., and Mitchekkm, J.E., 1977, Attitudes of Idaho off-road vehicle users and mangers: Moscow, Idaho, College of Forestry, Wildlife and Range Sciences, University of Idaho, Technical Report, 30 p.

Radforth, J.R., 1978, Higher travel speeds for off-road logging vehicles: Pointe-Claire, Quebec, Forest Engineering Research Institute of Canada, Report no. TR–22, 23 p.

Reed, P.C., and Haas, G.E., 1989, Off highway vehicles in Colorado—Estimated recreational use and expenditures: Fort Collins, Colorado, Colorado State University, Department of Recreation Resources and Landscape Architecture, Technical Report, 15 p.

Rhoads, W.A., Cochrane, S.A., and Williams, M.P., 1979, Status of endangered and threatened plant species on Tonopah Test Range—A survey: Goleta, California, Santa Barbara Research Center, Report no. SAND807035, 126 p.

Rickard, C.A., McLachlan, A., and Kerley, G.I.H., 1994, The effects of vehicular and pedestrian traffic on dune vegetation in South Africa: Ocean and Coastal Management, v. 23, no. 3, p. 225–247.

Rocky Mountain Research Initiative, 2002, Off-road vehicles in Colorado—Facts, trends, recommendations: Nederland, Colorado, Rocky Mountain Research Initiative, *http://www.rmri.org/rmri_full_report.pdf.*

Ross, J.B., and Willison, J.H.M., 1991, Impacts of all-terrain vehicles on bogs of the Cape Breton Highlands, Nova Scotia, Canada, Science and the management of protected areas— proceedings of an international conference held at Acadia University, Nova Scotia, Canada, 14-19 May 1991: Woldville, Nova Scotia, Canada, Acadia University, p. 533–534.

Schuett, M.A, and Ostergren, D., 2003, Environmental concern and involvement of individuals in selected voluntary associations: Journal of Environmental Education, v. 34, no. 4., p. 30–38.

Sheridan, D., 1979, Off-road vehicles on public land: Council on Environmental Quality, Report no. PB86211158, 94 p.

Shields, F.D., Cooper, C.M., Knight, S.S., Jr., and Moore, M.T., 2003, Stream corridor restoration research—A long and winding road: Ecological Engineering, v. 20, no. 5, p. 441–454.

Shindler, B., Brunson, M.W., and Cheek, K.A., 2004, Social acceptability in forest and range management, *in* Manfredo, M.J., Vaske, J.J., Bruyere, B.L., Field, D.R., and Brown, P.J., eds., Society and natural resources—A summary of knowledge: Jefferson, Missouri, Modern Litho, p. 146–157.

Siddharthan, R.V., Sebaaly, P.E., El-Desouky, M., Strand, D., and Huft, D., 2005, Heavy off-road vehicle tire–pavement interactions and response: Journal of Transportation Engineering, v. 131, no. 3, p. 239–247.

Sorenson, S., Kalivoda, M., Vacarro, R., Trozzi, C., and Samaras, Z., 1997, Data structure for estimating emissions from non-road sources: Lyngby, Denmark, Technical University of Denmark, Lyngby Institute, Report no. DTUETEO9702, 73 p.

Steiner, A.J., and Leatherman, S.P., 1981, Recreational impacts on the distribution of ghost crabs *Ocypode quadrata* Fab: Biological Conservation, v. 20, no. 2, p. 111–122.

Stephenson, G., 1999, Vehicle impacts on the biota of sandy beaches and coastal dunes—A review from a New Zealand perspective: Science for Conservation, v. 121, p. 1–14.

Stohlgren, T.J., and Parsons, D.J., 1992, Evaluating wilderness recreational opportunities—Application of an impact matrix: Environmental Management, v. 16, no. 3, p. 397–403.

Stokowski, P.A., and LaPointe, C.B., 2000, Environmental and social effects of ATVs and ORVs—An annotated bibliography and research assessment: Burlington, Vermont, University of Vermont, School of Natural Resources, 32 p.

Stynes, D.J., Nelson, C.M., and Lynch, J.A., 1988, State and regional economic impacts of snowmobiling in Michigan: East Lansing, Michigan, Department of Park, Recreation and Tourism Resources, Michigan State University, Technical Report, 31 p.

Swim, F.F., 1971, Bicycling and bicycle trails. A trails and trail based activities bibliography: U.S. Office of Library Services, Report no. BIB24, 38 p.

Turton, S.M., 2005, Managing environmental impacts of recreation and tourism in rainforests of the wet tropics of Queensland World Heritage Area: Geographical Research, v. 43, no. 2, p. 140–151.

Twiss, R., Sidener, J., Bingham, G., Burke, J.E., and Hall, C.H., 1980, Potential impacts of geothermal development on outdoor recreational use of the Salton Sea: Livermore, California, University of California, Livermore, Technical Report, 61 p.

U.S. Army Corps of Engineers, 1998, McGregor range land withdrawal, Fort Bliss, Otera County, New Mexico: U.S. Army Corps of Engineers, Report no. 980439, 586 p.

U.S. Army Corps of Engineers, 2001, Fort Bliss mission and master plan—Dona Ana and Otero Counties, New Mexico, and El Paso County, Texas: U.S. Army Corps of Engineers, Report no. 010081.

U.S. Bureau of Land Management, 1984, Proposed Monument resource management plan, Idaho: *U.S. Bureau of Land Management,*, Report no. 840584, 419 p.

U.S. Bureau of Land Management, 1985, Box Elder resource management plan, Utah: *U.S. Bureau of Land Management,*, Report no. 850448.

U.S. Bureau of Land Management, 1985, Grand Junction resource area—Resource management plan and environmental impact statement: *U.S. Bureau of Land Management,*, Technical Report, 171 p.

U.S. Bureau of Land Management, 1985, Medicine Lodge Proposed resource management plan, Bingham, Bonneville, Butte, Clark, Fremont, Jefferson, Madison, and Teton Counties, Idaho: *U.S. Bureau of Land Management,*, Report no. 850237.

U.S. Bureau of Land Management, 1985, Proposed Two Rivers resource management plan, Oregon: *U.S. Bureau of Land Management,*, Report no. 850409.

U.S. Bureau of Land Management, 1985, Two Rivers resource management plan, Oregon: *U.S. Bureau of Land Management,*, Report no. 850137, 163 p.

U.S. Bureau of Land Management, 1985, Yuma District resource management plan, Yuma, La Paz, and Mohave Counties, Arizona, and San Bernardino, Riverside, and Imperial Counties, California: *U.S. Bureau of Land Management,*, Report no. 850358, 310 p.

U.S. Bureau of Land Management, 1986, Baker resource management plan, Baker County, Oregon: *U.S. Bureau of Land Management,*, Report no. 860150, 129 p.

U.S. Bureau of Land Management, 1986, Lander resource management plan, Lander, Wyoming: *U.S. Bureau of Land Management,*, Report no. 860449.

U.S. Bureau of Land Management, 1986, Proposed 1985 amendments to the California desert conservation area plan and the Eastern San Diego County master framework plan, California: *U.S. Bureau of Land Management,*, Report no. 860432, 138 p.

U.S. Bureau of Land Management, 1986, Proposed 1985 amendments to the California desert conservation area plan, California: *U.S. Bureau of Land Management,*, Report no. 860077.

U.S. Bureau of Land Management, 1986, Proposed Baker resource management plan, Vale District, Oregon: *U.S. Bureau of Land Management,*, Report no. 860396.

U.S. Bureau of Land Management, 1986, resource management plan for the House Range resource area, Juab and Millard Counties, Utah: *U.S. Bureau of Land Management,*, Report no. 860377.

U.S. Bureau of Land Management, 1986, Utah statewide wilderness environmental impact statement: *U.S. Bureau of Land Management,*, Report no. 860036.

U.S. Bureau of Land Management, 1986, Wilderness suitability EIS—Grass Creek and Cody resource areas, Bighorn Basin, Wyoming (draft supplement for the Owl Creek wilderness study area): *U.S. Bureau of Land Management,*, Report no. 860168, 40 p.

U.S. Bureau of Land Management, 1987, Glenwood Springs resource area resource management plan, Colorado—Wilderness recommendations: *U.S. Bureau of Land Management,*, Report no. 870421, 168 p.

U.S. Bureau of Land Management, 1987, Medicine Bow-Divide resource areas resource management plan, Rawlins District, Wyoming: *U.S. Bureau of Land Management,*, Report no. 870212.

U.S. Bureau of Land Management, 1987, North Dakota resource management plan, Dunn and Bowman Counties, North Dakota: *U.S. Bureau of Land Management,*, Report no. 870236, 199 p.

U.S. Bureau of Land Management, 1987, Pinedale resource area resource management plan, Sublette and Lincoln Counties, Wyoming: *U.S. Bureau of Land Management,*, Report no. 870077, 347 p.

U.S. Bureau of Land Management, 1987, Pinedale resource area resource management plan, Sublette, Teton, Lincoln, and Fremont Counties, Wyoming: *U.S. Bureau of Land Management,*, Report no. 870428, 232 p.

U.S. Bureau of Land Management, 1987, Preliminary wilderness recommendations for the Arcata resource area, Eden Valley and Thatcher Ridge wilderness study areas, Ukiah District, Mendocino County, California: *U.S. Bureau of Land Management,*, Report no. 870348, 111 p.

U.S. Bureau of Land Management, 1987, Proposed Monument resource management plan and Wilderness Study, Idaho: *U.S. Bureau of Land Management,*, Report no. 870248, 162 p.

U.S. Bureau of Land Management, 1987, Proposed resource management plan and wilderness recommendations for the Lahontan resource area, Nevada: *U.S. Bureau of Land Management,*, Report no. 870293, 282 p.

U.S. Bureau of Land Management, 1987, Proposed utility corridor resource management plan, Alaska: *U.S. Bureau of Land Management,*, Report no. 870291, 281 p.

U.S. Bureau of Land Management, 1987, Proposed wilderness program for the Stafford District wilderness EIS Area, Cochise, Gila, Graham, and Greenlee Counties, Arizona and Hidalgo County, New Mexico: *U.S. Bureau of Land Management,*, Report no. 870156, 502 p.

U.S. Bureau of Land Management, 1987, Proposed wilderness program for the Upper Sonoran wilderness EIS Area, Maricopa, Mohave, Yavapai, and Yuma Counties, Arizona: *U.S. Bureau of Land Management,*, Report no. 870292, 394 p.

U.S. Bureau of Land Management, 1987, resource management plan for the Shoshone-Eureka resource area, Nevada—Wilderness recommendations: *U.S. Bureau of Land Management,*, Report no. 870403, 101 p.

U.S. Bureau of Land Management, 1987, Taos resource management plan, New Mexico: *U.S. Bureau of Land Management,*, Report no. 870331.

U.S. Bureau of Land Management, 1987, Washakie resource area resource management plan, portions of Big Horn, Hot Springs, and Washakie Counties, Wyoming: *U.S. Bureau of Land Management,*, Report no. 870395.

U.S. Bureau of Land Management, 1987, Wells resource area wilderness recommendations, Bluebell, Goshute Peak, South Pequop, and Bad Lands Wilderness Study Areas, Elko County, Nevada: *U.S. Bureau of Land Management,*, Report no. 870267, 224 p.

U.S. Bureau of Land Management, 1987, Wilderness recommendations for Clark County, Nevada: *U.S. Bureau of Land Management,*, Report no. 870108, 346 p.

U.S. Bureau of Land Management, 1987, Wilderness recommendations for the northcentral California wilderness study areas: *U.S. Bureau of Land Management,*, Report no. 870058, 115 p.

U.S. Bureau of Land Management, 1988, Brothers/Lapine resource management plan, Oregon: *U.S. Bureau of Land Management,*, Report no. 880365.

U.S. Bureau of Land Management, 1988, Central California section 202 wilderness study areas wilderness recommendations—Sheep Ridge, Milk Ranch/Case Mountain, and Ventana contiguous WSA'S: *U.S. Bureau of Land Management,*, Report no. 880020, 110 p.

U.S. Bureau of Land Management, 1988, Challis resource area proposed resource management plan, upper Columbia-Salmon Clearwater Districts, Custer and Lemhi Counties, Idaho: *U.S. Bureau of Land Management,*, Report no. 980508.

U.S. Bureau of Land Management, 1988, Cody resource area resource management plan, Big Horn and Park Counties, Wyoming: *U.S. Bureau of Land Management,*, Report no. 880315.

U.S. Bureau of Land Management, 1988, Fort Greely national maneuver area and Fort Greely air drop zone, Alaska—resource management plan: *U.S. Bureau of Land Management,*, Report no. 880276, 138 p.

U.S. Bureau of Land Management, 1988, Fort Wainwright maneuver area, Fairbanks North Star Borough, Alaska: *U.S. Bureau of Land Management,*, Report no. 880277, 138 p.

U.S. Bureau of Land Management, 1988, Pony Express resource management plan, Toole, Utah, and Salt Lake Counties, Utah: *U.S. Bureau of Land Management,*, Report no. 880310, 147 p.

U.S. Bureau of Land Management, 1988, Proposed Las Vegas resource management plan and final environmental impact statement, Clark and Nye Counties, Nevada: *U.S. Bureau of Land Management,*, Report no. 980207.

U.S. Bureau of Land Management, 1988, Uncompahgre Basin resource management plan, Colorado: *U.S. Bureau of Land Management,*, Report no. 880345.

U.S. Bureau of Land Management, 1988, Wilderness study areas in the Rock Springs District, Fremont, Lincoln, Sublette, and Sweetwater Counties, Wyoming (revised Draft environmental impact statement): *U.S. Bureau of Land Management,*, Report no. 880359, 175 p.

U.S. Bureau of Land Management, 1989, Arcata resource management plan, California: *U.S. Bureau of Land Management,*, Report no. 890306, 354 p.

U.S. Bureau of Land Management, 1989, Dixie resource area management plan, Washington County, Utah: *U.S. Bureau of Land Management,*, Report no. 890290, 250 p.

U.S. Bureau of Land Management, 1989, San Pedro River riparian management plan for the San Pedro River EIS area, Cochise County, Arizona: *U.S. Bureau of Land Management,*, Report no. 890152, 381 p.

U.S. Bureau of Land Management, 1989, San Rafael resource management plan, Emery County, Utah: *U.S. Bureau of Land Management,*, Report no. 890240.

U.S. Bureau of Land Management, 1989, Utility corridor proposed resource management plan, Alaska: *U.S. Bureau of Land Management,*, Report no. 890326, 281 p.

U.S. Bureau of Land Management, 1989, White Sands resource management plan amendment for McGregor Range, Otero County, New Mexico: *U.S. Bureau of Land Management,*, Report no. 890127.

U.S. Bureau of Land Management, 1990, Bishop resource management plan and environmental impact statement: *U.S. Bureau of Land Management,*, Report no. BLMCAES90001 1610, 273 p.

U.S. Bureau of Land Management, 1990, Rock Springs District wilderness study areas, Fremont, Lincoln, Sublette, and Sweetwater Counties, Wyoming: *U.S. Bureau of Land Management,*, Report no. 900439, 324 p.

U.S. Bureau of Land Management, 1991, Arizona Strip District resource management plan, Mohave and Coconino Counties, Arizona: *U.S. Bureau of Land Management,*, Report no. 910055, 617 p.

U.S. Bureau of Land Management, 1991, Safford District resource management plan, Arizona: *U.S. Bureau of Land Management,*, Report no. 910314, 511 p.

U.S. Bureau of Land Management, 1991, Spokane resource management plan, Washington: *U.S. Bureau of Land Management,*, Report no. 910374, 169 p.

U.S. Bureau of Land Management, 1991, Three Rivers resource management plan, Harney, Grant, Lake, and Malheur Counties, Oregon: *U.S. Bureau of Land Management,*, Report no. 910335.

U.S. Bureau of Land Management, 1992, Proposed Spokane resource management plan amendment and final environmental impact statement: *U.S. Bureau of Land Management,*, Report no. 920286, 181 p.

U.S. Bureau of Land Management, 1994, Fort Wainwright, Yukon maneuver area, proposed resource management plan and final environmental impact statement, Fairbanks North Star Borough, Alaska: *U.S. Bureau of Land Management,*, Report no. 940001, 124 p.

U.S. Bureau of Land Management, 1994, resource management plan and environmental impact statement for the Grass Creek resource area, Worland District, Wyoming: *U.S. Bureau of Land Management,*, Technical Report, 299 p.

U.S. Bureau of Land Management, 1994, Salem District resource management plan and environmental impact statement—Benton, Clackamas, Clatsop, Columbia, Lane, Lincoln, Linn, Marion, Multnomah, Polk, Tillamook, Washington, and Yamhill Counties, Oregon: *U.S. Bureau of Land Management,*, Report no. 940461.

U.S. Bureau of Land Management, 1994, Stateline resource management plan, Clark and Nye Counties, Nevada (draft environmental impact statement and draft supplement): *U.S. Bureau of Land Management,*, Report no. 940177.

U.S. Bureau of Land Management, 1997, Roswell and Carlsbad resource areas—Chaves, Curry, Debaca, Eddy, Guadalupe, Lea, Lincoln, Quay, and Roosevelt Counties, New Mexico: *U.S. Bureau of Land Management,*, Report no. 970072.

U.S. Bureau of Land Management, 1998, Judith-Valley-Phillips resource management plan—Chouteau, Fergus, Judith Basin, Petroleum, Phillips, and Valley Counties, Montana (draft supplement to the final environmental impact statement of September 1997): *U.S. Bureau of Land Management,*, Report no. 980149, 51 p.

U.S. Bureau of Land Management, 1998, Socorro resource area resource management plan, New Mexico: *U.S. Bureau of Land Management,*, Report no. 880397, 266 p.

U.S. Bureau of Land Management, 2000, Federal fluid minerals leasing and development, Otero and Sierra Counties, New Mexico: *U.S. Bureau of Land Management,*, Report no. 000383, 521 p.

U.S. Bureau of Land Management, 2003, Farmington resource management plan, San Juan, McKinley, Rio Arriba, and Sandoval Counties, New Mexico: *U.S. Bureau of Land Management,*, Report no. 030143.

U.S. Bureau of Land Management, 2004, Dillon resource management plan, Beaverhead and Madison Counties, Montana: *U.S. Bureau of Land Management,*, Report no. 040153, 626 p.

U.S. Bureau of Land Management, 2005, Newmont Mining Company, Emigrant project, Elko County, Nevada: *U.S. Bureau of Land Management,*, Report no. 050117, 269 p.

U.S. Bureau of Land Management, 2006, Coeur d'Alene resource management plan, Idaho: *U.S. Bureau of Land Management,*, Report no. 060002, 347 p.

U.S. Bureau of Land Management and U.S. Forest Service, 1986, Proposed Coronado National Forest land and resource management plan, Arizona and New Mexico: *U.S. Bureau of Land Management* and U.S. Forest Service, Report no. 860307.

U.S. Bureau of Land Management and U.S. Forest Service, 2004, North Fork of the South Platte and the South Platte Rivers wild and scenic river study report, Pike and San Isabel National Forests and Comanche and Cimarron National Grasslands, Douglas, Jefferson, Park, and Teller Counties, Colorado: U.S. Bureau of Land Management and *U.S. Forest Service,* Report no. 040037, 301 p.

U.S. Bureau of Reclamation, 2001, Pothole Reservoir resource management plan, Grant County, Washington: U.S. Bureau of Reclamation, Report no. 010518, 567 p.

U.S. Department of the Air Force, 1992, Defense evaluation support activity testing and evaluation program, Kirtland Air Force Base, New Mexico: *U.S. Air Force,* Report no. 920325, 487 p.

U.S. Federal Highway Administration, 2005, US 33 Nelsonville bypass, city of Nelsonville, Hocking and Athens Counties, Ohio: *U.S. Federal Highway Administration*, Report no. 050282, 828 p.

U.S. Fish and Wildlife Service, 1985, Alaska Peninsula National Wildlife Refuge final comprehensive conservation plan, environmental impact statement, and wilderness review: *U.S. Fish and Wildlife Service,* Technical Report, 426 p.

U.S. Fish and Wildlife Service, 1986, Great Dismal Swamp National Wildlife Refuge master plan, Virginia and North Carolina: *U.S. Fish and Wildlife Service,* Report no. 860501, 245 p.

U.S. Forest Service, 1973, Management of South Holston unit: *U.S. Forest Service,* Report no. USDAFSFESADM7350, 121 p.

U.S. Forest Service, 1985, Kisatchie National Forest land and resource management plan, Louisiana: *U.S. Forest Service,* Report no. 850414.

U.S. Forest Service, 1985, Proposed Land and resource management plan for the Medicine Bow National Forest and Thunder Basin National Grassland, Wyoming: *U.S. Forest Service,* Report no. 850523.

U.S. Forest Service, 1986, Chequamegon National Forest, Wisconsin land and resource management plan: *U.S. Forest Service,* Report no. 860325.

U.S. Forest Service, 1986, Cleveland National Forest land and resource management plan, Orange, Riverside, and San Diego Counties, California: *U.S. Forest Service,* Report no. 860209.

U.S. Forest Service, 1986, Croatan and Uwharrie National Forests land and resource management plan, North Carolina: *U.S. Forest Service,* Report no. 860218.

U.S. Forest Service, 1986, Environmental impact statement for the Coronado National Forest plan: *U.S. Forest Service,* Technical Report, 275 p.

U.S. Forest Service, 1986, Environmental impact statement, Gila National Forest plan: *U.S. Forest Service,* Technical Report, 347 p.

U.S. Forest Service, 1986, Land and resource management plan, Chippewa National Forest, Beltrami, Cass, and Itasca Counties, Minnesota: *U.S. Forest Service,* Report no. 860208.

U.S. Forest Service, 1986, National Forests in Alabama land and resource management plan: *U.S. Forest Service,* Report no. 860095.

U.S. Forest Service, 1986, Proposed land and resource management plan for Wayne National Forest, Ohio: *U.S. Forest Service,* Report no. 860426.

U.S. Forest Service, 1986, Proposed Santa Fe National Forest land and resource management plan, Mora, San Miguel, Santa Fe, Sandoval, Los Alamos, and Rio Arriba Counties, New Mexico: *U.S. Forest Service,* Report no. 860013.

U.S. Forest Service, 1987, Environmental impact statement for the Apache-Sitgreaves National Forests Plan: *U.S. Forest Service,* Technical Report, 392 p.

U.S. Forest Service, 1987, Environmental impact statement, Santa Fe National Forest plan: *U.S. Forest Service,* Technical Report, 366 p.

U.S. Forest Service, 1987, Proposed Apache-Sitgreaves National Forests land and resource management plan, Apache, Coconino, Greenlee, and Navajo Counties, Arizona: *U.S. Forest Service,* Report no. 870387.

U.S. Forest Service, 1987, Proposed land and resource management plan for the Coconino Forest, Coconino, Gila, and Yavapai Counties, Arizona: *U.S. Forest Service,* Report no. 870288.

U.S. Forest Service, 1987, Santa Fe National Forest land management plan, Mora, San Miguel, Santa Fe, Sandoval, Los Alamos, and Rio Arriba Counties, New Mexico: *U.S. Forest Service,* Report no. 870299.

U.S. Forest Service, 1988, Proposed Kaibab National Forest land and resource management plan, Coconino, Yavapai, and Mohave Counties, Arizona: *U.S. Forest Service,* Report no. 880109.

U.S. Forest Service, 1988, Record of decision for USDA, Forest Service—Final environmental impact statement, Allegheny National Forest, Land and resource management plan—Elk, Forest, McKean, and Warren Counties, Pennsylvania: *U.S. Forest Service,* Technical Report, 41 p.

U.S. Forest Service, 1989, San Bernardino National Forest land and resource management plan, San Bernardino and Riverside Counties, California: *U.S. Forest Service,* Report no. 890023.

U.S. Forest Service, 1990, Cherokee National Forest land and resource management plan, Tennessee, North Carolina, and Virginia (final supplement to the final environmental impact statement of April 1986): *U.S. Forest Service,* Report no. 900062, 62 p.

U.S. Forest Service, 1990, Umatilla National Forest land and resource management plan, Oregon and Washington: *U.S. Forest Service,* Report no. 900237.

U.S. Forest Service, 1991, Corral Mountain timber sale, San Juan National Forest, Archuleta County, Colorado: *U.S. Forest Service,* Report no. 910124, 59 p.

U.S. Forest Service, 1992, Martinez Creek timber sale, San Juan National Forest, Colorado: *U.S. Forest Service,* Report no. 920275, 75 p.

U.S. Forest Service, 1992, Sierra National Forest land and resource management plan, Fresno, Madera, and Mariposa Counties, California: *U.S. Forest Service,* Report no. 920105.

U.S. Forest Service, 1993, Six Rivers National Forest plan, Humboldt, Del Norte, Siskiyou, and Trinity Counties, California: *U.S. Forest Service,* Report no. 930340.

U.S. Forest Service, 1994, Oregon Dunes National Recreation Area management plan, Siuslaw National Forest—Coos, Douglas, and Lane Counties, Oregon: *U.S. Forest Service,* Report no. 940292.

U.S. Forest Service, 1994, Revised land and resource management plan for the national forests and grasslands in Texas—Angelina, Fannin, Houston, Jasper, Montague, Montgomery, Nacogdoches, Newton, Sabine, San Augustine, San Jacinto, Shelby, Trinity, Walker, and Wise Counties, Texas: *U.S. Forest Service,* Report no. 940385.

U.S. Forest Service, 1995, Draft environmental impact statement, Fish Bate analysis area—North Fork Ranger District, Clearwater National Forest, Clearwater County, Idaho: *U.S. Forest Service,* Technical Report.

U.S. Forest Service, 1995, East Fork Blacks Fork analysis area, Wasatch-Cache National Forest, Summit County, Utah (final supplement to the final environmental impact statement of July 1992): *U.S. Forest Service,* Report no. 950294, 49 p.

U.S. Forest Service, 1995, Mendocino National Forest land and resource management plan—Colusa, Glenn, Lake, Mendocino, Tehama, and Trinity Counties, California: *U.S. Forest Service,* Report no. 950337.

U.S. Forest Service, 1996, Proposed revised land and resource management plan for Francis Marion National Forest, Berkeley and Charleston Counties, South Carolina: *U.S. Forest Service,* Report no. 960158.

U.S. Forest Service, 2000, Silver Creek integrated resource project, Emmett Ranger District, Boise National Forest, Boise and Valley Counties, Idaho: *U.S. Forest Service,* Report no. 000241, 264 p.

U.S. Forest Service, 2002, Sixshooter project, Emmett Ranger District, Boise National Forest, Gem County, Idaho: *U.S. Forest Service,* Report no. 020238, 201 p.

U.S. Forest Service, 2002, Uncompahgre National Forest Travel Plan Revision, Gunnison, Hinsdale, Mesa, Montrose, Ouray, San Juan, and San Miguel Counties, Colorado (final supplement to the final environmental impact statement of April 2000): *U.S. Forest Service,* Report no. 020131, 92 p.

U.S. Forest Service, 2002, Whiskey Campo Resource Management Project, Mountain Home Ranger District, Boise National Forest, Elmore County, Idaho: *U.S. Forest Service,* Report no. 020299, 151 p.

U.S. Forest Service, 2003, Big Bend Ridge vegetation management project and timber sale, Ashton/Island Park Ranger District, Caribou-Targhee National Forest, Freemont County, Idaho: *U.S. Forest Service,* Report no. 030225.

U.S. Forest Service, 2003, Cross-country travel by off-highway vehicles, Apache-Sitgreaves, Coconino, Kaibab, Prescott, and Tonto National Forests, Arizona: *U.S. Forest Service,* Report no. 030184, 235 p.

U.S. Forest Service, 2003, Duck Creek–Swains access management project, Cedar City Ranger District, Dixie National Forest, Iron, Garfield, and Kane Counties, Utah: *U.S. Forest Service,* Report no. 030343, 49 p.

U.S. Forest Service, 2004, Duck Creek fuels treatment analysis, Cedar City Ranger District, Dixie National Forest, Kane County, Utah: *U.S. Forest Service,* Report no. 040535, 240 p.

U.S. Forest Service, 2004, French Face ecosystem restoration, Ninemile Ranger District, Lolo National Forest, Montana: *U.S. Forest Service,* Report no. 040336, 184 p.

U.S. Forest Service, 2005, Dean project area, Bearlodge Ranger District, Black Hills National Forest, Crook County, Wyoming: *U.S. Forest Service,* Report no. 050212, 192 p.

U.S. Forest Service, 2005, Final environmental impact statement for the access designation on the Ocala National Forest, Lake, Marion, and Putnam Counties, Florida: *U.S. Forest Service,* Technical Report.

U.S. Forest Service, 2005, Gallatin National Forest travel management plan, Gallatin, Madison, Park, Meagher, Sweetgrass, and Carbon Counties, Montana: *U.S. Forest Service,* Report no. 050231.

U.S. Forest Service, 2005, Rocky Mountain Ranger District travel management plan, Rocky Mountain Ranger District, Lewis and Clark National Forest, Glacier, Pondera, Teton, and Lewis and Clark Counties, Montana: *U.S. Forest Service,* Report no. 050236, 386 p.

U.S. Forest Service, 2005, Woodrock project, Tongue Ranger District, Bighorn National Forest, Sheridan County, Wyoming: *U.S. Forest Service,* Report no. 050112, 201 p.

U.S. Forest Service, 2006, Hoosier National Forest land and resource management plan—Final environmental impact statement: *U.S. Forest Service,* Technical Report.

U.S. Forest Service, U.S. Bureau of Land Management, and U.S. Fish and Wildlife Service, 1997, George Washington National Forest land and resource management plan, Virginia and West Virginia (final supplement to the final environmental impact statement of January 1993): *U.S. Forest Service,* U.S. Bureau of Land Management, and U.S. Fish and Wildlife Service, Report no. 970046.

U.S. National Park Service, 1998, Redwood National and State Parks general management plan, Del Norte and Humboldt Counties, California: *U.S. National Park Service,* Report no. 980290, 474 p.

U.S. National Park Service, 1999, Big Cypress National Preserve, recreational off-road vehicle management plan—Collier, Miami-Dade, and Monroe Counties, Florida (draft supplement to the environmental impact statement of October 1991): *U.S. National Park Service,* Report no. 990285, 171 p.

U.S. National Park Service, 2000, Recreational off-road vehicle management plan, Big Cypress National Preserve—Collier, Miami-Dade, and Monroe Counties, Florida (final supplement to the final environmental impact statement of October 1991): *U.S. National Park Service,* Report no. 000275, 619 p.

Vancini, F.W., 1989, *Policy and management considerations for off-road vehicles—Environmental and social impacts*: Ithaca, New York, Cornell University, 19 p.

Watson, A., Asp, C., Walsh, J. and Kulla, A., 1997, *The contribution of research to managing conflict among national forest users*: Trends, v. 34, no. 3, p. 29–3 5.

Webb, R.H., and Wilshire, H.G., 1983, *Environmental effects of off-road vehicles—Impacts and management in arid regions:* New York, Springer-Verlag, 534 p.

Weeden, R., 1976, *On consumptive users—A myth: Alaska Conservation Review*, v. 17, no. 3, p. 15.

Wilshire, H.G., Nakata, J.K., S*hipley, Susan, and Prestegaard, Karen, 1978, Impacts of vehicles on natural terrain at seven sites in the San Francisco Bay area: Environmental Geology,* v. 2, no. 5, p. 295–3 19.

Wilshire, H.G., Shipley, Susan, and Nakata, J.K., 1978, Impacts of off-road vehicles on vegetation: *Transactions of the North American Wildlife and Natural Resources Conference,* v. 43, p. 131–139.

Wolcott, T.G., and Wolcott, D.L., 1984, Impact of off-road vehicles on macroinvertebrates of a mid-Atlantic beach: *Biological Conservation,* v. 29, no. 3, p. 2 17–240.

Appendix 2. Search Methods and Results of Off-Highway Vehicle Effects Literature and Internet Resources

2.1. Methods

2.1.1. Literature Search

From May 10–26, 2006, a comprehensive literature search was conducted to encapsulate the current knowledge on effects of off-highway vehicle (OHV) activities as it pertains to natural resource attributes addressed by the Bureau of Land Management's (BLM) land health standards (U.S. Bureau of Land Management, 2001; Pellant and others, 2005). The search was conducted through 33 electronic literature databases available at Colorado State University's (CSU) library (table 2.1) and search engines available on the Internet. The databases searched encompassed all major, and some minor, sources of relevant literature, including professional, peer-reviewed journal papers and technical reports/articles published in magazines representing the industrial sector and non-governmental conservation organizations.

Search terms used included "OHV," "off-highway vehicle," "ORV," or "off-road vehicle" combined with terms representing the BLM's land health standards, including soil health and watershed condition, nutrient cycling, wildlife health and habitat quality for native plants and animals (especially for species of special status), water quality, and air quality (table 2.1). In addition, searches were conducted on the socioeconomics of OHV use. Within each database searched, all search terms were applied to the 26 topic areas (see footnote 1 associated with table 2.1) listed in CSU's library database subject list. Relevant citations also were gleaned from highly relevant reports and journal articles. All relevant citations identified were grouped by land-health categories to build extensive bibliographies for each land-health category. If a given citation was relevant to more than one land-health category, it was listed in each of the related bibliographies.

Table 2.1. Search terms used and publication years included when using search engines and 33 electronic literature databases at Colorado State University's library to assemble an extensive bibliography of literature on effects of off-highway vehicles

Database searched (for publications years)1	Search term										
	Air quality	Bene-fits	Domes-tic live-stock grazing	Erosion	Noise	Soil impacts	Travel and transportation management	Visual impa-cts	Vegetation impacts	Water quality	Wildlife impacts
AGRICOLA EBSCO (1970-2006)	X		X	X		X			X	X	X
AGRICOLA National Agricultural Library (1970-2006)	X		X	X		X			X	X	X
Agricultural and Environmental Biotechnology Abstracts (1993-2006)	X			X	X	X	X		X	X	X
Agricultural Sciences (in Cambridge Scientific Abstracts) (compilation of multiple databases2)	X		X	X		X			X	X	X
Animal Behavior Abstracts (1982-2006)		X	X		X						X
Applied Science and Technology	X			X	X	X	X	X		X	

Abstracts (1983-2006)											
ASFA 1: Biological Sciences and Living Resources (1971-2006)					X				X		X
ASFA 2: Ocean Technology, Policy & Non-Living Resources (1971-2006)	X			X		X	X	X			
ASFA 3: Aquatic Pollution and Environmental Quality (1990-2006)	X			X	X	X	X	X		X	
BioEngineering Abstracts (1993-2006)				X		X	X				
Biological Abstracts (1969-2006)			X						X		X
Biological Sciences (1982-2006)			X						X		X
Biological Sciences (in Cambridge Scientific Abstracts) (compilation of multiple databases2)			X						X		X

Table 2.1. Continued

Database searched (for publications years)1	Search term										
	Air quality	Bene-fits	Domes-tic live-stock grazing	Erosion	Noise	Soil impacts	Travel and transportation management	Visual impa-cts	Vegetation impacts	Water quality	Wildlife impacts
Biology Digest (1989-2006)									X		X
Biotechnology and Bioengineering Abstracts (1982-2006)				X		X	X				
CAB Abstracts Archive (1900-1973)											X
Ecology Abstracts (1969-2006)		X	X						X	X	X
EIS: Digests of environmental impact statements (1985-2006)							X				
Engineering Index (1884-2006)				X		X	X				
Environmental Engineering Abstracts (1990-2006)				X		X	X				
Environmental Sciences & Pollution Management	X			X	X	X	X				

(compilation of multiple databases2)											
Forest Service database (dates not indicated)			X	X		X	X		X	X	X
Geobase (1980-2006)				X		X					
Human Population & Natural Resource Management (1995-2006)			X				X				
Plant Management Network (compilation of multiple databases2)									X		
Plant Science (1994-2006)									X		
Pollution Abstracts (1981-2006)	X				X		X				
Pollution Management & Environmental Sciences (in Cambridge Scientific Abstracts) (compilation of multiple databases2)	X				X		X				

Table 2.1. Continued

Database searched (for publications years)1	Search term										
	Air quality	Bene-fits	Domes-tic live-stock grazing	Erosion	Noise	Soil impacts	Travel and transportation management	Visual impa-cts	Vegetation impacts	Water quality	Wildlife impacts
Threatened and Endangered Species System (dates not indicated)									X	X	X
Toxicology Abstracts (1981-2006)	X					X					
Water Resources Abstracts (1967-2006)										X	
Wildlife & Ecology Studies Worldwide (1935-2006)											X
Zoological Record (1978-2006)											X

2.1.2. Internet Search

Specific goals of the Internet search were to (1) identify websites and other Internet resources provided by the BLM and U.S. Forest Service (FS) regions 2 (time constraints limited this search to Colorado), 3, and 4; (2) classify OHV-related Internet websites by focus/intent and resource type; and (3) report on highly relevant websites. Two primary products were subsequently developed: (1) a thesaurus (table 2.2) of search topics and terms related to OHV effects, and (2) a list of significant BLM and FS resources and corresponding Internet websites. The thesaurus was developed to identify Internet-based resources regarding OHVs and the types of OHV effects. All possible combinations of search terms were used, but any one search string depended on the search engine used. Although *FirstGov* is a good search engine for finding government publications and reports, it limits search results to 100. In contrast, *Google* returns up to 1000 results, and its filter may be used to constrain searches to a specific type of website (for example, ".gov"); however, *Google* is unable to employ multiple exclusion criteria to exclude irrelevant websites (for example, it cannot specify "NOT FS" and NOT BLM") whereas *FirstGov* does handle multiple exclusion criteria. In other words, *Google* was searched with an OHV term combined with an impact term while limiting any one search to "BLM," "EPA" (U.S. Environmental Protection Agency), "FS" (U.S. Forest Service), "NPS" (U.S. National Park Service), or "USGS" (U.S. Geological Survey) sites. Then *FirstGov* was searched with the same terms while excluding "BLM," "EPA," "FS," "NPS," "USGS," and ".com" domains, which captured websites representing individual States, military agencies, and other non-commercial entities.

Table 2.2. Search topics and their associated search terms used in searching the Internet for publications and documents pertaining to off-highway vehicle effects and policies

Search topic	Search term(s)
Off-highway vehicle	OHV Off-road vehicle Off-road vehicles
Air quality	Air quality
Benefits	Benefits
Domestic livestock grazing	Domestic livestock grazing
Erosion	Erosion
Human dimension	Human dimension Human dimensions
Noise	Noise
Soil impact	Soil impact Soil impacts
Transportation management	Transportation management
Travel	Travel
Vegetation impact	Vegetation impact Vegetation impacts

Visual impact	Visual impact Visual impacts
Water quality	Water quality
Wildlife impact	Wildlife impact Wildlife impacts

After the Internet searches were conducted, each website identified were reviewed (limited to 2 minutes per website) and assigned a relevance-class code (relevance to OHV effects and policies; table 2.3). Websites classified as H (highly relevant), R (relevant), or S (slightly relevant) were further categorized by focus area (table 2.4). All BLM and FS regions 2 (Colorado only), 3, and 4 websites were included in the results tallies; websites containing news releases, specific flora assessments, and BLM Resource Advisory Council meetings were omitted. Highly relevant websites were selected for presentation.

2.2. Results

2.2.1. Literature Resources

The literature search, and additional sources uncovered outside of the formal search, yielded approximately 700 citations, a number of which overlap in terms of their relevance to categories of land health (table 2.5).

Table 2.3. Relevance class codes and definitions pertaining to Internet websites found to contain information regarding off-highway vehicle effects and policies

Relevance class	Definition
H	Highly relevant—high-quality resources related directly to OHV impacts;
	generally included number of OHV trail miles and visitor days, or reasons why OHVs excluded from specific locations; includes OHV strategy/plan documents or public communication sites on OHV recreation areas
R	Relevant—targeted mention of OHV but not as specific or detailed as a highly relevant site
S	Slightly relevant—mention of OHV but with few supporting statements or details
U	Unrelated—no or minimal OHV content
Z	Unable to access, page would not load, URL has changed, page loads but is empty of content
ZDup	Content duplicate of previous entry

Table 2.4. Focus areas and definitions of Internet websites found to contain information regarding off-highway vehicle effects and policies

Focus area	Definition
Administration	Information sourced from the Federal level
Citizen input	Site contains or informs about public comments on OHV use
EIS	Site contains a draft or final environmental impact statement (EIS) orenvironmental assessment (EA)—attempt was to keep this category to sites related to actions up through completion of a management plan although there is some overlap with the next category due to title ambiguity
Impact	Site focus was measurement of impact
Legal	Response to appeals regarding travel management plans or EIS documents
Manual	Site containing a handbook or manual
Management plan	Sites with completed management plans or actions stemming from implementation of a management plan; includes monitoring, revision of plans, assessment for roadless areas, and road analyses
Monitoring	Usually annual reports of monitoring activity prescribed by a management plan
Press	Site with information in the form of a press release or media announcement
Road guide	Sites intended to provide public information about OHV and recreation site use; includes descriptions, trail maps, event calendars, safety, licensing and regulation, and closure information

Table 2.5. Number of relevant publications found, and publication dates included, in a literature search on effects of off-highway vehicle activity as they pertain to the U.S. Bureau of Land Management's land health standards

Land health category	*No. citations*	*Years included*
Soils	314	1959-2006
Air quality	104	1970-2006
Water quality	218	1959-2005
Vegetation	326	1962-2006
Wildlife	387	1970-2006
Land users	211	1967-2006

2.2.2. Internet Resources

The Internet search yielded nearly 30,000 State and Federal government websites, of which 8,693 were unique (a single HTML page or .pdf file); 23 percent (n = 1,998) were BLM websites, 55 percent (n = 4,817) were FS websites, and 7 percent (n = 568) were NPS websites (table 2.6). FS regions 2 (Colorado only), 3, and 4 together represented 12 percent of the websites. Of the 8,693 unique sites, 2,495 were visited and reviewed (29 percent) using the methods described above. All search term combinations returned at least 100 results, and some returned as many as 3,700 (table 2.6).

Of the unique websites identified, only 13% ($n = 335$) were classified as highly relevant; 15 percent were relevant, 22 percent were slightly relevant, and 50 percent were unrelated/not available (figure 2.1). Forty-seven percent of those classified as highly relevant were BLM sites, and 53 percent were FS sites. The majority of highly relevant sites (68 percent) were dedicated to environmental impact statements (EIS) or management plans (figure 2.2). Only two percent of the highly relevant sites focused on measuring or monitoring the effects of OHV activities. When all sites were considered, a slightly higher percentage of sites (approximately 5 percent) included monitoring or impact assessment (figure 2.3).

Key similarities between highly relevant BLM and FS websites (figures 2.4 and 2.5) included equal emphasis on road guides (about 12 percent of sites) and little emphasis on monitoring and impacts (0 and 1 percent for BLM, respectively; 1 and 1 percent for FS, respectively). The BLM sites had slightly greater emphasis on EISs and management plans (73 percent) than did FS sites (65 percent). Interestingly, 10 percent of the FS sites concerned legal issues—primarily appeals to decisions—whereas no material of this type was found on BLM websites. This may be due to an agency decision of what type of material is posted to the Internet or it may be a result of a difference in EIS/management plan implementation status between the two agencies. Table 2.8 presents examples of five highly ranked web sites (if available) from BLM and/or FS Region 2 for each of the focus areas.

Table 2.6. Search results, by topic, for all Internet websites pertaining to off-highway vehicle effects and policies (n = 22,990)

Search topic	Results
Air quality	3,040
Benefits	3,081
Domestic livestock grazing	373
Erosion	3,228
Human dimension(s)	241
Noise	2,650
Soil impact(s)	461
Transportation management	453
Travel	3,726
Vegetation impact(s)	155
Visual impact(s)	1,392

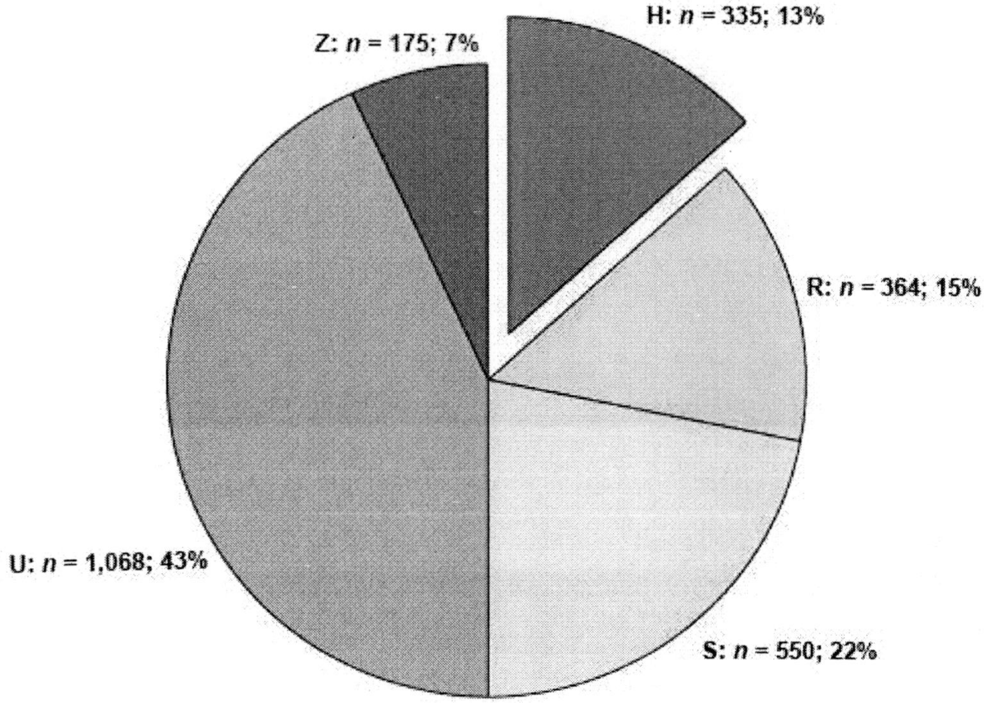

Figure 2.1. Breakdown of unique Internet websites (n = 2,495) classified as "highly relevant" (H), "relevant" (R), "slightly relevant" (S), and "unrelated/unavailable" (U, Z) to off-highway vehicle effects and policies.

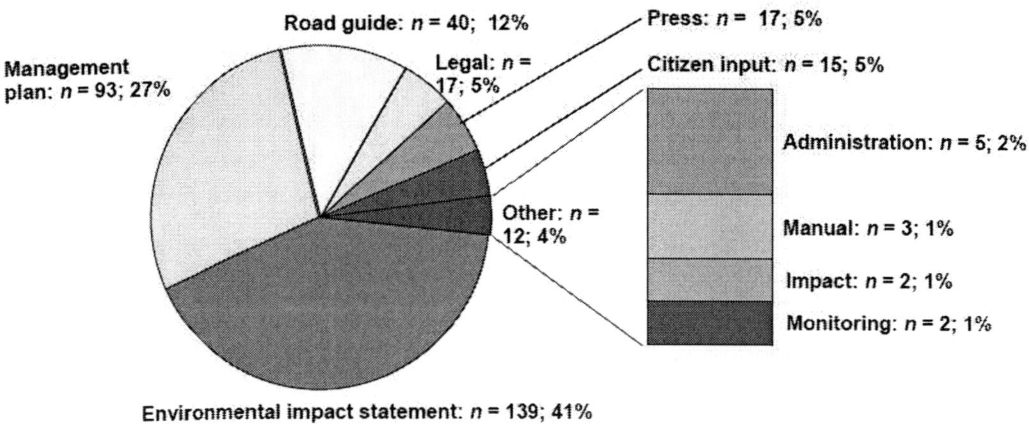

Figure 2.2. Focus areas of Internet websites (n = 333) classified as highly relevant to off-highway vehicle effects and policies.

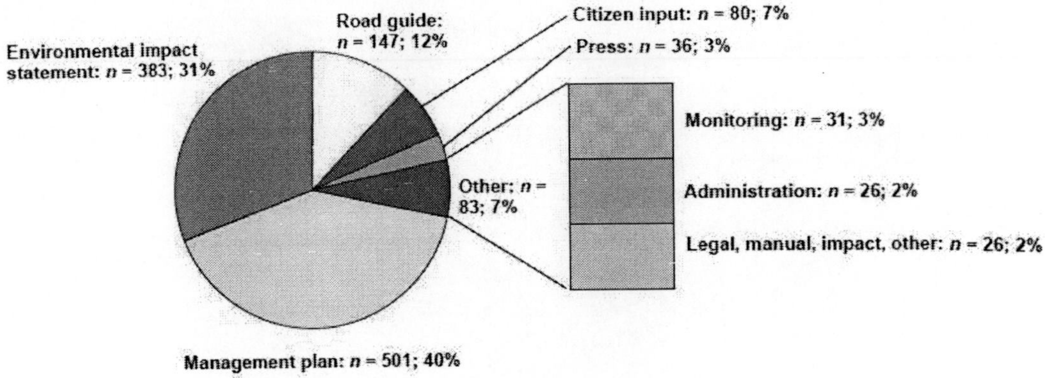

Figure 2.3. Focus areas for all Internet websites (n = 1,230) classified as highly relevant, relevant, and somewhat relevant to off-highway vehicle effects and policies.

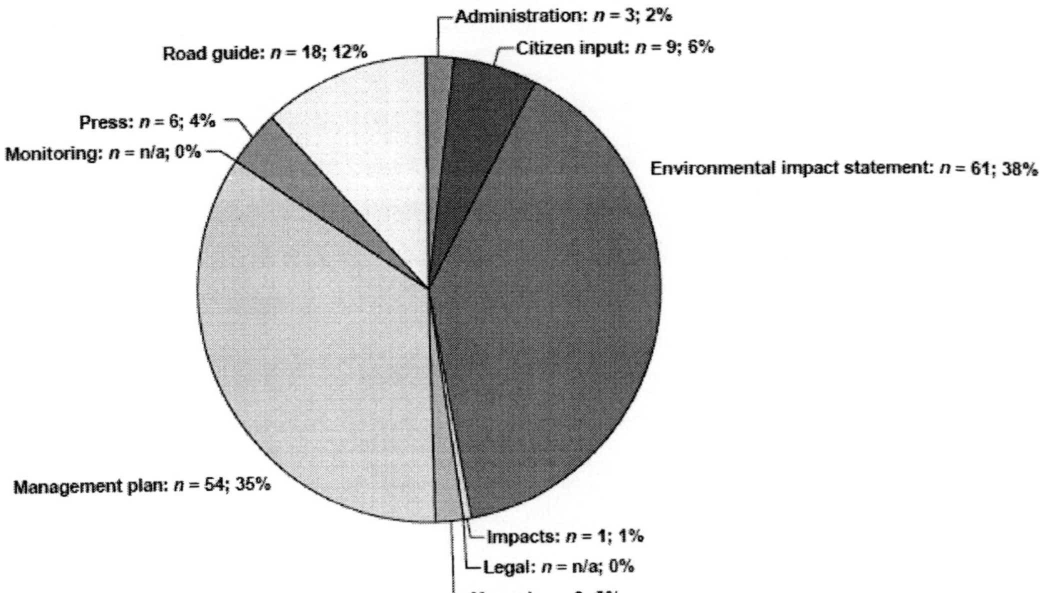

Figure 2.4. Focus areas of U.S. Bureau of Land Management Internet websites (n = 155) classified as highly relevant, relevant to off-highway vehicle effects and policies.

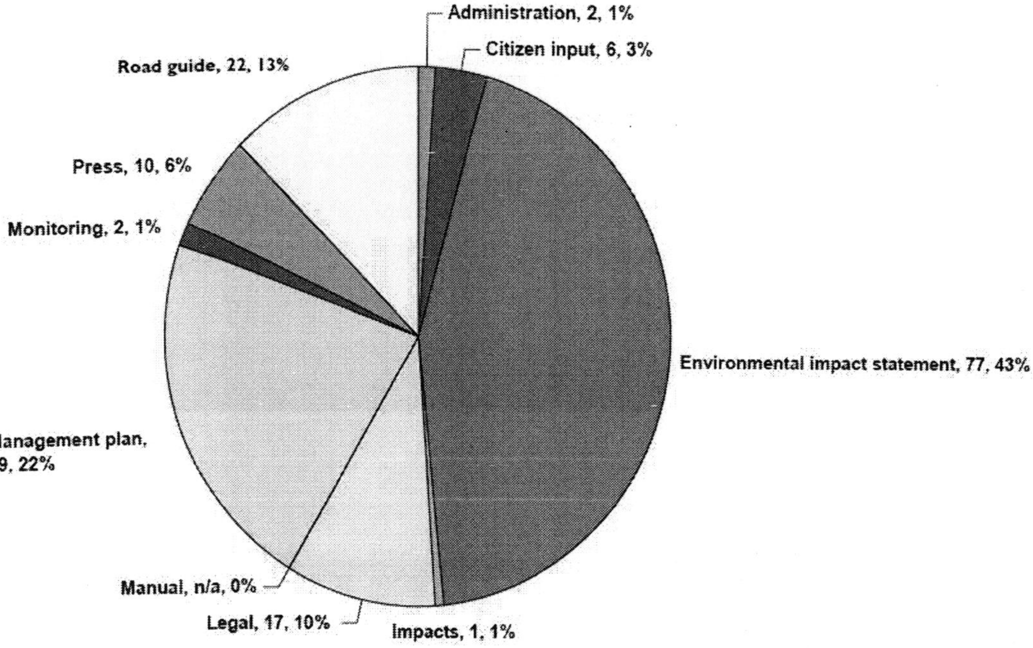

Figure 2.5. Focus areas of U.S. Forest Service (regions 2 [Colorado only], 3, and 4 only) Internet websites (n = 176) classified as highly relevant to off-highway vehicle effects and policies.

INDEX

A

B

C

D

E

F

G

H

I

M

N

O

P

Q

R

S

T

U

V

W

X

Y

Z